Dragons
Chinese Teaching Specialists

# Secondary School Chinese *for GCSE*

## Workbook 2

***Secondary School Chinese for GCSE Workbook 2*** has been written to accompany AQA GCSE Chinese Textbook 2. The contents of this workbook are fully aligned to the units of the textbook, and give students valuable opportunities to practise what they have learned.

Each unit contains a wealth of activities, covering reading, writing, grammar and culture. The book's special focus on character writing and knowledge of correct tones will enable students to communicate effectively and correctly. Reading activities range from simple sentences to long passages which introduce supplementary vocabulary.

Our hope is that this book will be invaluable for both teacher and student: for the teacher, it gives a wide selection of activities that can be used in class or for homework. These activities are carefully designed to match the level of the student. For the student, it will give opportunity for review and consolidation on any topic on which he/she needs extra help.

# Unit 9 Section 1 Vocabulary

| | | |
|---|---|---|
| 单人房 | dānrénfáng | single room |
| 双人房 | shuāngrénfáng | double room |
| 登记表 | dēngjì biǎo | registration form |
| 身份证 | shēnfènzhèng | ID card |
| 护照 | hùzhào | passport |
| 旅行 | lǚxíng | travel, to travel |
| 旅行袋 | lǚxíngdài | suitcase |
| 丢 | diū | to lose |
| 拿 | ná | to carry, take |
| 旅游车 | lǚyóuchē | tourist coach |
| 目的地 | mùdìdì | destination |
| 出发 | chūfā | to depart, set off |
| 到达 | dàodá | to arrive |
| 单程 | dānchéng | single (ticket) |
| 来回 | láihuí | return (ticket) |
| 天安门 | Tiān'ānmén | The Gate of Heavenly Peace |
| 广场 | guǎngchǎng | square |
| 鸟巢 | Niǎocháo | The Bird's Nest Stadium |
| 站台 | zhàntái | platform |
| 安全 | ānquán | safety |
| 安全带 | ānquándài | seat belt |
| 酒店 | jiǔdiàn | large hotel |
| 免费 | miǎnfèi | free (of charge) |
| 故宫 | Gùgōng | The Forbidden City |
| 换钱 | huàn qián | to exchange currency |
| 接 | jiē | to receive; to collect somebody after travel |
| 送 | sòng | to give as a gift; to see somebody off |

**Online vocabulary**

| | | |
|---|---|---|
| 明信片 | míngxìnpiàn | postcard |
| 远 | yuǎn | far |
| 离 | lí | (particle for indicating distance) |
| 近 | jìn | near |
| 参观 | cānguān | to visit |
| 过 | guò | (particle to indicate experience) |
| 不得了 | bùdéliǎo | extremely, very |
| 麻烦 | máfan | trouble, burden |
| 烤鸭 | kǎoyā | roast duck |
| 快乐 | kuàilè | happy |
| 极了 | jíle | extremely, very |

# Unit 9 Section 1 Stroke Order

单人房
dānrénfáng

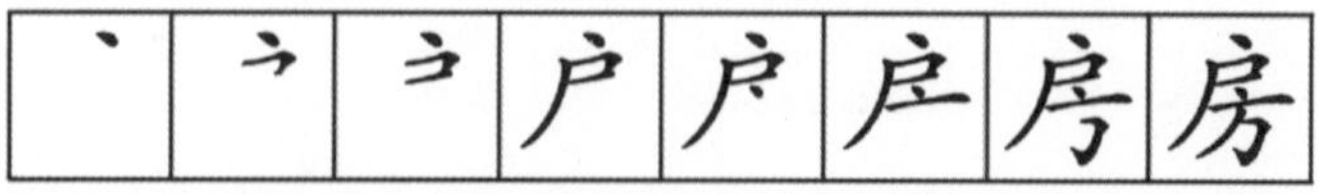

双人房
shuāngrénfáng

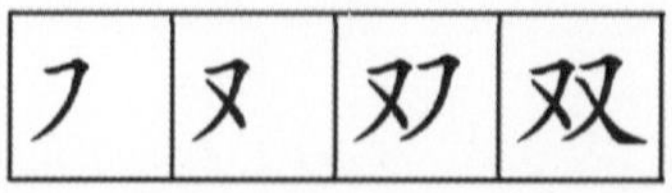

旅行
lǚxíng

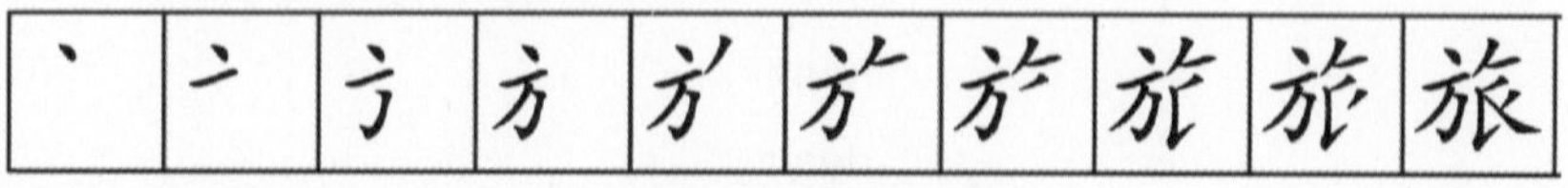

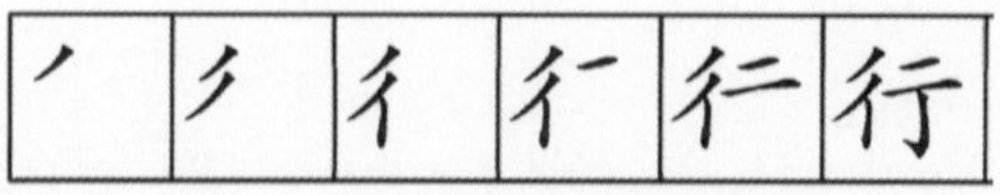

出发
chūfā

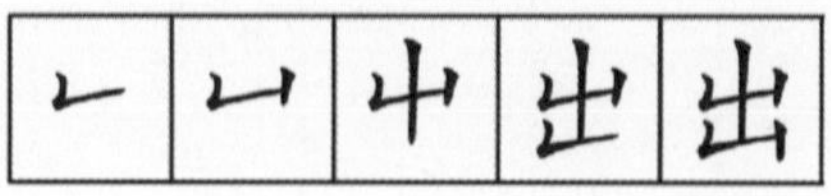

离
lí

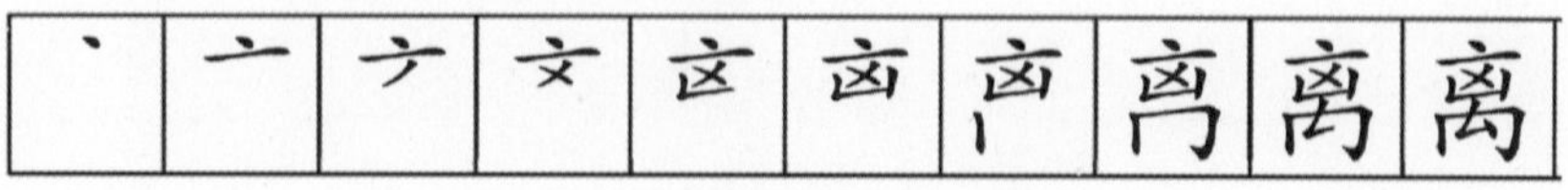

近
jìn

远
yuǎn

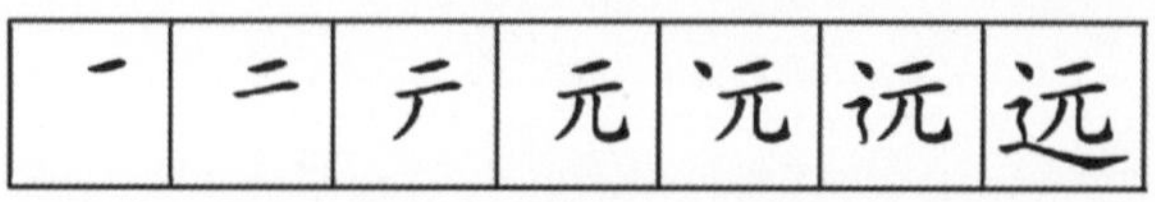

参观
cānguān

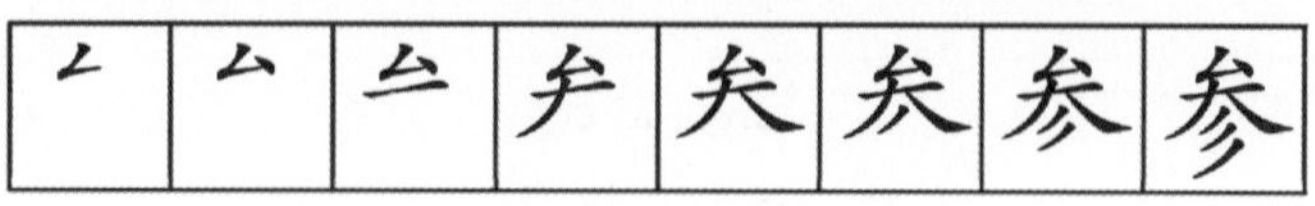

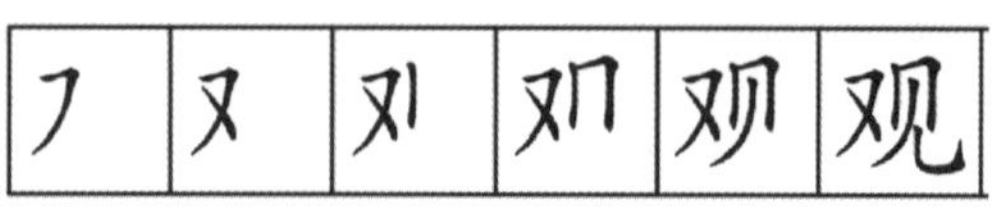

送
sòng

丢
diū

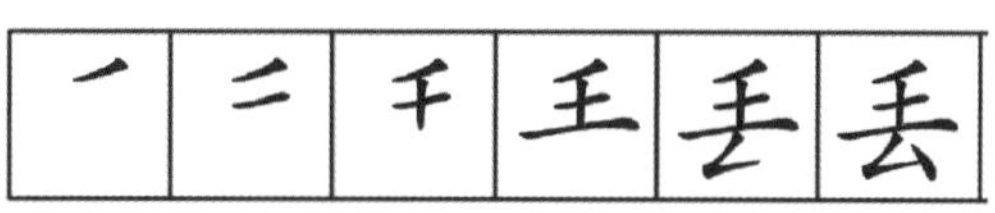

拿
ná

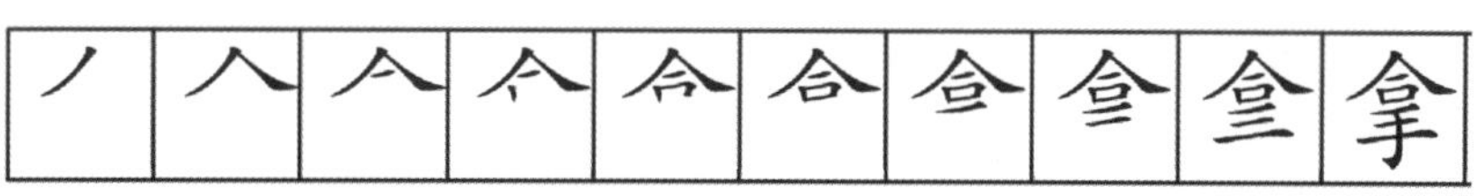

# 观光 Sightseeing

## Section 1 – On holiday

Activity 9.1

Practise writing the following characters:

单

房

双

旅

行

出

离

近

远

参

观

送

丢

拿

## Activity 9.2

Draw lines to match the following verbs and objects.

| | |
|---|---|
| 送 | 天安门广场 |
| 吃 | 目的地 |
| 丢 | 朋友 |
| 到达 | 护照 |
| 参观 | 烤鸭 |
| 订 | 双人房 |

## Activity 9.3

Split up the following words and write each under the correct heading according to the tone of each character. As an example, 换 has been completed for you.

换钱　出发　旅游车　快乐　安全带

| 1st tone | 2nd tone |
|---|---|
| | |

| 3rd tone | 4th tone |
|---|---|
| | 换 |

## Activity 9.4

Draw a line to match each Chinese word with its English translation.

| | |
|---|---|
| 身份证 | single room |
| 明信片 | platform |
| 站台 | one-way ticket |
| 旅行袋 | ID card |
| 单程票 | registration form |
| 单人房 | postcard |
| 登记表 | suitcase |

## Activity 9.5

Write the pinyin and tones of these sentences.

1. 你家离银行近不近？
2. 他做的饺子好吃极了！
3. 请问，北京酒店在哪儿？
4. 买飞机票很麻烦！
5. 你快去超市吧，有很多东西是免费的！

## Activity 9.6

Below, fill in the blanks with appropriate words.

1. 邮局离公园不____。
2. 我____见过他，不知道他是____。
3. 我弟弟七点____达北京，我去火车站接他。
4. 我不____你去汽车站，你自己去，好不好？
5. 你只有美元；你要不要换____？

美元 Měi Yuán = US dollar

没　送　钱　远　谁　到

### Activity 9.7
Translate the following sentences into Chinese.

1. May I see your ID?
2. I've lost it!
3. Let's visit the Bird's Nest today!
4. The train will set off at 6:00pm.
5. Your passport photo doesn't look too good.
6. I'll help you carry your suitcase, okay?

### Activity 9.8
Match the start and end of each sentence to make a logical complete sentence.

| | |
|---|---|
| 请给我写 | 站台等她。 |
| 坐旅游车 | 明信片，好吗？ |
| 我们在大酒店 | 护照。 |
| 我们坐在 | 住了三天。 |
| 请您拿好 | 很便宜，也很舒服！ |

### Activity 9.9
Below are some places around town. Write a sentence for each to express whether the place is near or far from your home.

1. 市中心
2. 你的学校
3. 你最喜欢的饭馆

### Activity 9.9 (continued)

4. 你最喜欢的商店
5. 火车站

### Activity 9.10
Read the passage about Dominic's visit to Beijing and answer the questions in English.

我以前没来过北京。昨天飞机到达北京以后，我坐地铁去了酒店。我睡了两个小时的觉，下午四点出去走了走。因为天安门广场离酒店很近，所以我去了那儿参观。我饿了，在路上吃了饺子。好吃极了！明天我想吃北京烤鸭。

1. How many times has Dominic been to Beijing?
2. What did he do when he got to his hotel?
3. What did he do at 4pm?
4. What did he eat for dinner?
5. What does he want to do tomorrow?

## Activity 9.11

Look at the booking screen below and answer the questions. Use complete Chinese sentences.

◉ 单程 ○ 来回 出发地 西安 目的地 上海 出发日 2020-03-20 查询

| 车次 | 出发站 | 到达站 | 出发时间 | 到达时间 | 历时 | 商务座 | 一等座 | 二等座 | 无座 |
|---|---|---|---|---|---|---|---|---|---|
| G1914 | 西安北 | 上海 | 06:26 | 13:22 | 06:56 | 16 ¥2114.0 | 有 ¥1095.0 | 有 ¥669.5 | -- |
| T206 | 西安北 | 上海 | 00:47 | 17:49 | 17:02 | -- | -- | -- | 无 ¥180.5 |
| G1972 | 西安北 | 上海 | 13:50 | 21:08 | 07:18 | 有 ¥2114.0 | 无 ¥1095.0 | 有 ¥669.5 | -- |

1. Where are you trying to go?
2. What is your city of departure?
3. What is the shortest journey?
4. How expensive is the cheapest train?
5. Are you buying a single or return ticket?

## Activity 9.12

Answer the questions about yourself in Chinese. Use complete sentences.

1. 你有哪国的护照？
2. 你最想参观什么地方？
3. 你觉得坐旅游车去旅行怎么样？
4. 你最不想丢的东西是什么？
5. 你每天上学几点出门？

## Activity 9.13
Translate the following passage into English.

我跟弟弟去南京旅行。我们叔叔住在南京的农村，可是他还是来市中心的火车站接我们。他也帮我们拿旅行袋，客气极了！因为叔叔的家不大，所以我和弟弟订了酒店的双人房。酒店离叔叔的家很近。南京天气不太好，热得不得了！我们参观了很漂亮的寺庙。

## Activity 9.14
Translate the following passage into Chinese.

I went to Shanghai with my friend. We took the train from Xi'an, setting off at 9pm on Friday. Because Shanghai is very far from Xi'an, and the travel time was long, we slept on the train. In Shanghai we visited The Bund. My friend lost his wallet so I bought him lunch.

# Unit 9 Section 2 Vocabulary

| | | |
|---|---|---|
| 纪念品 | jìniànpǐn | souvenirs |
| 信用卡 | xìnyòngkǎ | credit card |
| 现金 | xiànjīn | cash |
| 门票 | ménpiào | entrance ticket |
| 慢 | màn | slow |
| 爬山 | pá shān | to climb a mountain |
| 假期 | jiàqī | holiday |
| 天坛 | Tiān Tán | Temple of Heaven |
| 京剧 | jīngjù | Beijing opera |
| 晒太阳 | shài tàiyáng | to sunbathe, to tan |
| 博物馆 | bówùguǎn | museum |
| 度假 | dù jià | to spend a holiday |

(No online vocabulary)

# Unit 9 Section 2 Stroke Order

|  |  |  |
|---|---|---|
| 安全 | ānquán | 丿 人 亼 今 仐 全 |
| 酒店 | jiǔdiàn | 丶 冫 氵 汇? 汀? 沔? 洒? 洒? 酒? 酒 |
| 过 | guò | 一 十 寸 寸 讨 过 |
| 极了 | jíle | 一 十 才 木 朳? 杈? 极 |
| 广场 | guǎngchǎng | 丶 亠 广 |
| 用 | yòng | 丿 冂 月 月 用 |
| 信用卡 | xìnyòngkǎ | 丿 亻 亻? 仁? 仁? 信? 信? 信? 信 |
| | | 丨 卜 上 卡? 卡 |
| 慢 | màn | 丨 忄? 忄? 忄 怛? 怛? 怛? 怛? 慍? 慍? 慍? 慍? 慢? 慢 |
| 京剧 | jīngjù | ㇇ ㇆? 尸 尸? 居? 居? 居 居 剧? 剧 |
| 度假 | dùjià | 丶 亠 广 产? 庐? 庐? 庐? 度? 度 |
| | | 丿 亻 亻? 仃? 仴? 仴? 仴? 假? 假? 假? 假 |
| 图书馆 | túshūguǎn | 丨 冂 冂? 冈? 冈? 图? 图? 图 |
| 回 | huí | 丨 冂 冂? 冋? 冋? 回 |
| 思 | sī | 丨 冂 日? 田? 田 田? 思? 思? 思 |

## Section 2 – Seeing the sights

Activity 9.16
Practise writing the following characters:

| | | | | | | | | | | | |
|---|---|---|---|---|---|---|---|---|---|---|---|
| 全 | | | | | | | | | | | |
| 酒 | | | | | | | | | | | |
| 过 | | | | | | | | | | | |
| 极 | | | | | | | | | | | |
| 广 | | | | | | | | | | | |
| 用 | | | | | | | | | | | |
| 信 | | | | | | | | | | | |
| 卡 | | | | | | | | | | | |
| 慢 | | | | | | | | | | | |
| 剧 | | | | | | | | | | | |
| 度 | | | | | | | | | | | |
| 假 | | | | | | | | | | | |
| 图 | | | | | | | | | | | |
| 回 | | | | | | | | | | | |
| 思 | | | | | | | | | | | |

### Activity 9.17
Draw lines to match each tourist attraction with its English name.

| | |
|---|---|
| 鸟巢 | Tian'anmen Square |
| 天安门广场 | The Great Wall |
| 天坛 | The Forbidden City |
| 故宫 | The Bird's Nest |
| 长城 | The Temple of Heaven |

### Activity 9.18
Split up the following words and write each under the correct heading according to the tone of each character. As an example, 京 has been completed for you.

京剧　　度假　　博物馆　　现金　　爬山

| 1st tone | 2nd tone |
|---|---|
| 京 | |
| **3rd tone** | **4th tone** |
| | |

### Activity 9.19
Draw a line to match each Chinese word with its English translation.

| | |
|---|---|
| 度假 | to sunbathe |
| 纪念品 | credit card |
| 晒太阳 | to spend one's holiday |
| 假期 | souvenirs |
| 信用卡 | trouble, burden |
| 麻烦 | holiday |

### Activity 9.20
Write the pinyin and tones of these sentences.

1. 这些纪念品都便宜极了！
2. 你的假期过得怎么样？
3. 爬山没有意思，我们去公园晒太阳吧！
4. 天坛离鸟巢远不远？怎么走？
5. 参观博物馆不用买门票。

### Activity 9.21
Below, fill in the blanks with appropriate words.

1. 你看过京＿吗？
2. 坐＿共汽车比＿自行车＿。
3. 你拿现金了吗？这里不能用＿＿＿。
4. 我最想＿＿天安门＿＿。
5. 我吃＿烤鸭，好吃＿了！

公　过　参观　慢　剧
广场　骑　极　信用卡

## Activity 9.22

Translate the following sentences into Chinese.

1. We had dinner in the hotel. It was ridiculously expensive!

2. They do T'ai Chi in the square every morning.

3. Please speak a little slower.

4. I'll see you tomorrow in the library.

5. My Chinese teacher went back to Beijing to spend the holiday.

6. He doesn't have a credit card. He's too young.

## Activity 9.23

Match the start and end of each sentence to make a logical complete sentence.

| | |
|---|---|
| 天安门广场 | 不老，是2008年建设的。 |
| 鸟巢 | 是大公园，里面有很大的寺庙。 |
| 以前，故宫 | 的风景非常好，有很多山。 |
| 长城 | 是皇帝的家。现在是博物馆了。 |
| 天坛 | 在故宫南边。 |

建设 jiànshè = to build

## Activity 9.24

You plan to stay with a Chinese friend but you don't want to trouble them. Below are some things that they might do to be a good host; using 不用, write a Chinese sentence for each to say there is no need, and why.

1. 给我做很多饭

2. 给我买面包、咖啡

3. 很早起床

4. 送我去汽车站

### Activity 9.25
Read the passage about Rashida's outing and answer the questions.

今天我参观了历史博物馆。我最喜欢看的是中国画。我一个人去的，所以一个人慢慢看。我在那儿看画看了两个半小时。因为博物馆的展览很有意思，所以我不想回家。可是博物馆下午四点半关门，我只好回酒店休息一下。晚上我去大市场买了新的旅行袋。

1. Where did Rashida go during the day?
2. Who accompanied her?
3. How long did she stay there?
4. Why did she leave?
5. What did she do in the evening?

### Activity 9.26
Answer the questions for yourself in Chinese. Use complete sentences.

1. 你用过信用卡吗？你买了什么东西？
2. 你想去哪儿度假？
3. 离你家很近的地方有没有广场？
4. 你在图书馆喜欢做什么？
5. 你旅行的时候会买什么纪念品？

### Activity 9.27
What would you say in each of the following situations?

1. You would like to pay for something but the till is unattended.
2. You are giving a presentation on Russia and want to survey the audience about who has been there.
3. You need a pen and would like to borrow one.
4. You are looking for your neighbour but nobody answers when you knock.
5. You need the toilet but are unsure whether it is occupied.

### Activity 9.28
Translate the following passage into English.

我非常喜欢爬山，风景很漂亮。爬山也不用买门票，是免费的活动。我知道妹妹不喜欢在外边玩儿。可是因为她没爬过山，所以我请她跟我一起去。她走得很慢。她还说我们走得太远。

### Activity 9.29
Translate the following passage into Chinese.

I really love to travel. I often go to little shops to look for postcards. Yesterday I bought a postcard for my grandad; it had a picture of the Great Wall on it. However, I can't write the postcard because I lost my pen. Tomorrow I will buy some new pens in the city centre.

# Unit 10 Section 1 Vocabulary

| | | |
|---|---|---|
| 耳朵 | ěrduo | ear |
| 鼻子 | bízi | nose |
| 腿 | tuǐ | leg |
| 手 | shǒu | hand |
| 脚 | jiǎo | foot, feet |
| 嘴 | zuǐ | mouth |
| 牙齿 | yáchǐ | tooth, teeth |
| 头 | tóu | head |
| 疼 | téng | to hurt, ache |
| 别……了 | bié...le | don't... |
| 怎么办? | zěnme bàn? | What to do? |
| 前天 | qiántiān | the day before yesterday |
| 看病 | kàn bìng | to see a doctor |
| 西药 | Xī yào | Western medicine |
| 中药 | Zhōng yào | Chinese medicine |
| 生病 | shēng bìng | to become ill |
| 病 | bìng | illness, disease; to become ill |
| 针灸 | zhēnjiǔ | acupuncture |
| 让 | ràng | to let, allow; to make (somebody do something) |
| 药 | yào | medicine |
| 伤 | shāng | to injure; a wound |
| 难 | nán | difficult |
| 感冒 | gǎnmào | to catch a cold |
| 上班 | shàng bān | to go to work |
| 身体 | shēntǐ | body |
| 健康 | jiànkāng | health, healthy |

**Online vocabulary**

| | | |
|---|---|---|
| 多久? | duōjiǔ? | how long? |
| 不要 | bú yào | don't (command) |
| 后天 | hòutiān | the day after tomorrow |
| 不用 | b bú yòng | no need, don't |

# Unit 10 Section 1 Stroke Order

健康
jiànkāng

耳朵
ěrduo

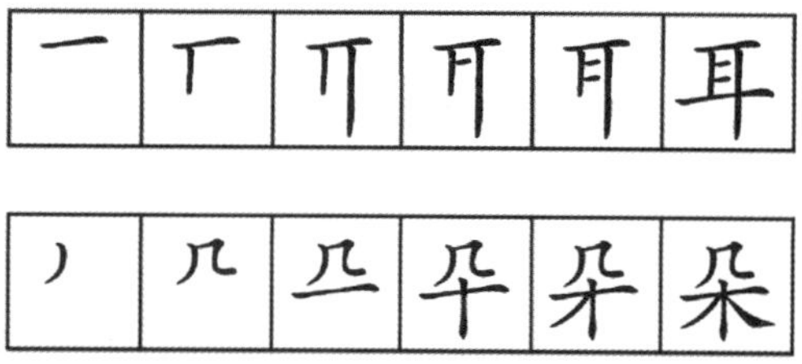

身体
shēntǐ

疼
téng

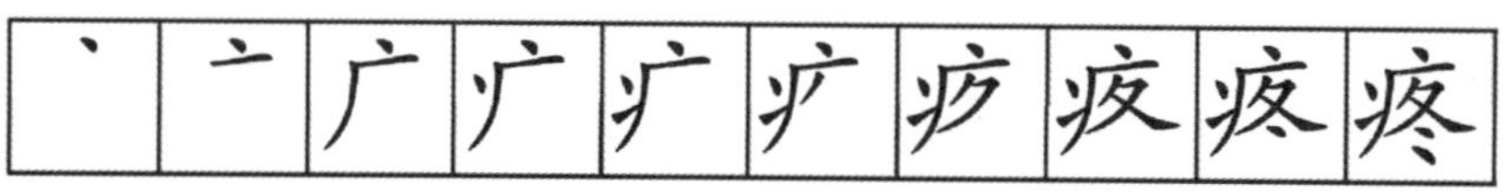

病
bìng

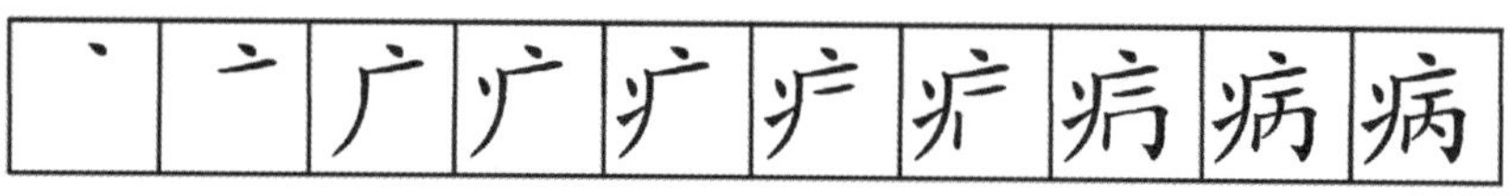

药
yào

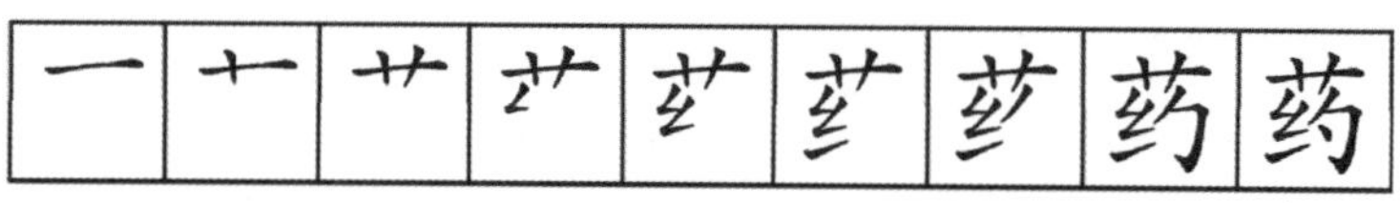

让
ràng

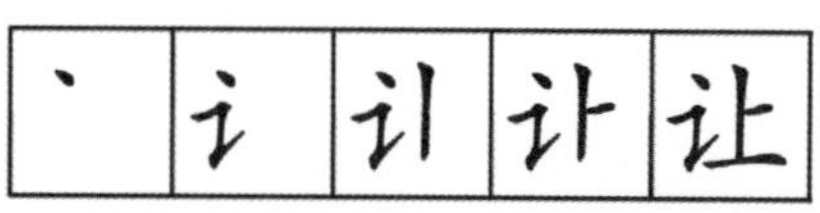

别
bié

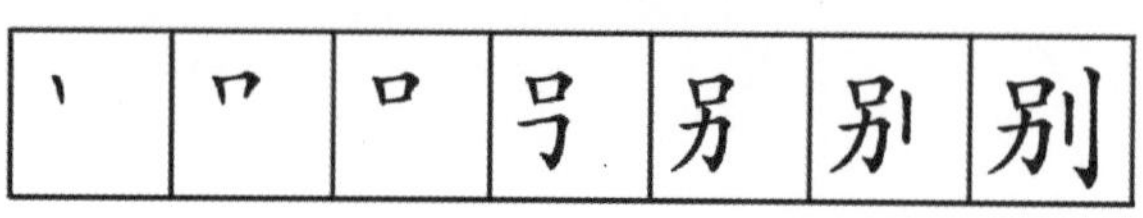

难
nán

上班
shàngbān

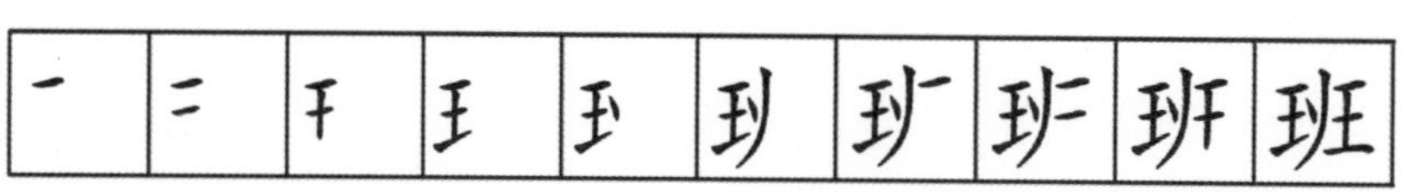

多久
duōjiǔ

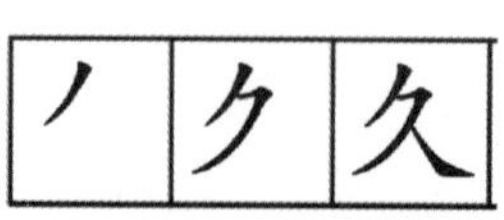

# 健康 Health

## Section 1 – Healthy Living

Activity 10.1

Practise writing the following characters:

| | | | | | | | | | | | |
|---|---|---|---|---|---|---|---|---|---|---|---|
| 健 | | | | | | | | | | | |
| 康 | | | | | | | | | | | |
| 耳 | | | | | | | | | | | |
| 朵 | | | | | | | | | | | |
| 身 | | | | | | | | | | | |
| 体 | | | | | | | | | | | |
| 疼 | | | | | | | | | | | |
| 病 | | | | | | | | | | | |
| 药 | | | | | | | | | | | |
| 让 | | | | | | | | | | | |
| 别 | | | | | | | | | | | |
| 难 | | | | | | | | | | | |
| 班 | | | | | | | | | | | |
| 久 | | | | | | | | | | | |

## Activity 10.2

Draw lines to match each body part to the corresponding function.

| | |
|---|---|
| 耳朵 | 看电视 |
| 眼睛 | 踢足球 |
| 牙齿 | 听音乐 |
| 脚 | 吃东西 |
| 手 | 拿东西 |

## Activity 10.3

Split up the following words and write each under the correct heading according to the tone of each character. As an example, 生 has been completed for you.

生病　感冒　身体　健康　西药

| 1st tone | 2nd tone |
|---|---|
| 生 | |
| **3rd tone** | **4th tone** |
| | |

## Activity 10.4

Draw a line to match each Chinese word with its English translation.

| | |
|---|---|
| 后天 | to become ill |
| 前天 | the day before yesterday |
| 生病 | acupuncture |
| 看病 | to visit the doctor |
| 针灸 | to ache, hurt |
| 疼 | the day after tomorrow |

## Activity 10.5

Write the pinyin and tones of these sentences.

1. 阿姨的鼻子很大！
2. 我的头已经疼了两天了。应该怎么办？
3. 我伤了我的手，很疼！
4. 你觉得西药还是中药好？
5. 天气冷了，要戴帽子。你不要感冒了！

## Activity 10.6

Below, fill in the blanks with appropriate words.

1. 医生让我多吃菜，可是很☐！我喜欢吃肉！
2. 请☐好☐份证，不☐丢！
3. 爷爷八十岁，可是他的身☐☐☐极了！
4. 你生☐了，去看病吧。医生会给你☐吃。
5. 你的鼻子疼不☐？

| | | | | |
|---|---|---|---|---|
| 要 | 药 | 拿 | 体 | 难 |
| 身 | 疼 | 病 | 健康 | |

## Activity 10.7

Translate the following sentences into Chinese.

1. The doctor made me take medicine.
2. She is very healthy! (Literally, 'Her body is healthy.')
3. How long have your ears hurt?
4. I am ill and cannot go to work today.
5. It is difficult to write because my hand hurts.

## Activity 10.8

Match the start and end of each sentence to make a logical complete sentence.

| | |
|---|---|
| 我的头疼， | 对你的身体很好。 |
| 你每天几点 | 可是我不想做。 |
| 多喝水 | 你要多穿衣服！ |
| 外边冷得不得了， | 上班？ |
| 医生让我做针灸， | 今天不想去看电影。 |

## Activity 10.9

Below are some examples of lifestyle habits. Write a sentence for each to state whether the habit is good or bad for your health.

1. 吃很多猪肉
2. 每天吃苹果

## Activity 10.9 (continued)

3. 跟小孩子玩儿
4. 每天喝六杯咖啡
5. 跑步

## Activity 10.10

Read the passage about Ari's trip to the doctor and answer the questions in English.

我前天和姐姐在公园玩儿。我伤了我的腿。姐姐送我去了医院看医生。医生说我不用吃药，他说我应该多休息。姐姐帮我走回家。爸爸看了我的腿，对我说"明天你不用去学校。我给老师打电话"

1. When did Ari injure himself?
2. What part of his body was injured?
3. Who took him to see the doctor?
4. What was the doctor's advice?
5. What was Ari's father's response?

### Activity 10.11

Answer the questions about yourself in Chinese. Use complete sentences.

1. 你做过针灸吗？
2. 你吃过中药吗？
3. 你生病了会做什么？
4. 朋友生病了，你会怎么办？
5. 你每天做什么保持身体健康？

保持 bǎochí = maintain

### Activity 10.12

The following are some common expressions using 别. Guess their meanings or look them up and write the English translations below.

1. 别着急！
2. 别担心！
3. 别提了！
4. 别忘了！
5. 别走了！
6. 别这样！

### Activity 10.13

Translate the following passage into English.

今天我奶奶感冒了，不能上班。她让我去超市买药。因为她是中国人，所以她要吃中药。她说吃了两天的药，身体就会好。她觉得不用去看病，她还说自己知道怎么办。

### Activity 10.14

Translate the following passage into Chinese.

Although I'm very healthy, my ears often hurt. Sometimes I see a doctor and she gives me medicine. But going to the doctor is difficult for me because I live far away. The medicine is also not free of charge. What can I do!

# Unit 10 Section 2 Vocabulary

| | | |
|---|---|---|
| 参加 | cānjiā | to attend |
| 米 | mǐ | metre |
| 跟……一样 | gēn...yīyàng | the same as... |
| 矮 | ǎi | short (in height) |
| 分 | fēn | minute |
| 更 | gèng | more (in comparison) |
| 快 | kuài | fast |
| 金牌 | jīnpái | gold medal |
| 赛车 | sàichē | car racing |
| 队员 | duìyuán | teammate |
| 体操 | tǐcāo | gymnastics |
| 比赛 | bǐsài | competition; to compete |
| 滑雪 | huá xuě | skiing; to ski |
| 滑冰 | huá bīng | skating; to skate |
| 哪年 | nǎ nián | which year? |
| 大家 | dàjiā | everyone |

**Online vocabulary**

| | | |
|---|---|---|
| 次 | cì | an occurrence, 'time' |
| 体育馆 | tǐyùguǎn | gymnasium, stadium |
| 健身房 | jiànshēnfáng | gym (with weight machines, etc) |
| 早餐 | zǎocān | breakfast |
| 午餐 | wǔcān | lunch |
| 晚餐 | wǎncān | dinner |
| 炒 | chǎo | to fry |
| 最近 | zuìjìn | recently |

# Unit 10 Section 2 Stroke Order

休息
xiūxi

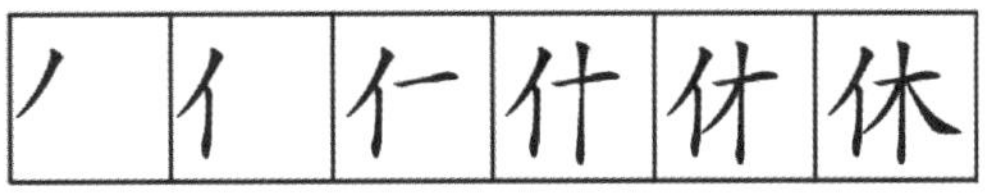

更
gèng

队员
duìyuán

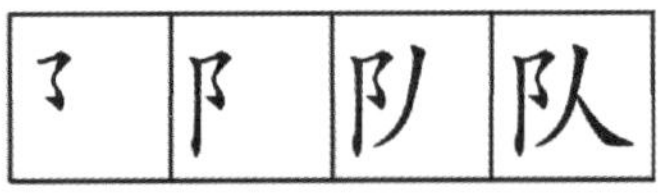

参加
cānjiā

比赛
bǐsài

滑冰
huábīng

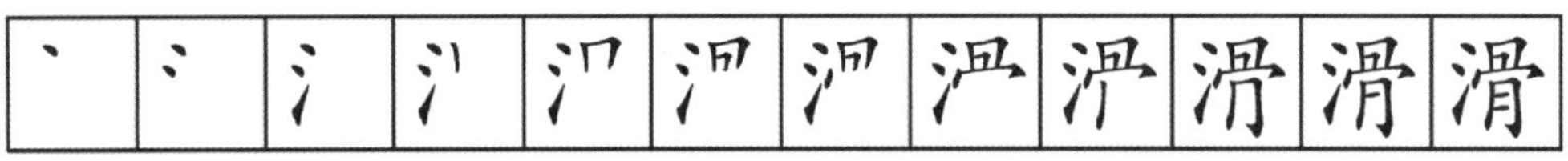

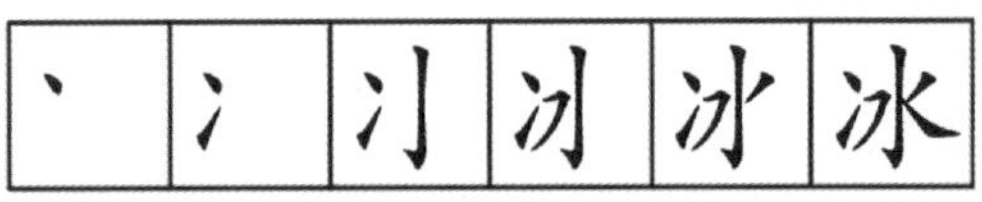

次
cì

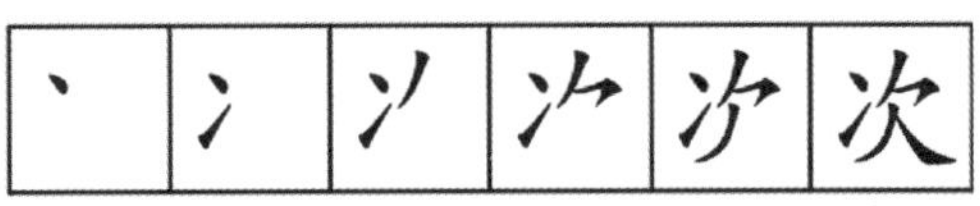

体育馆
tǐyùguǎn

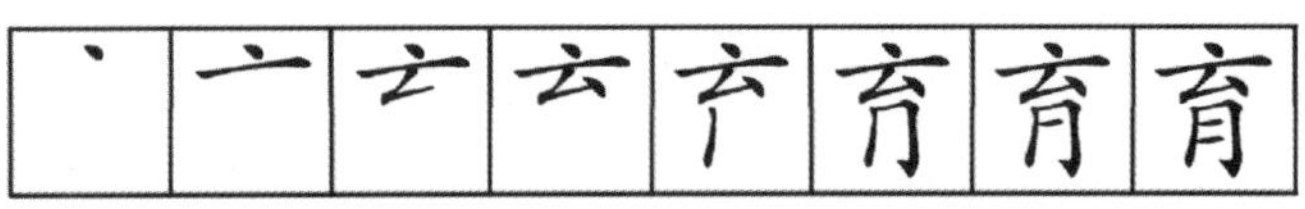

早餐
zǎocān

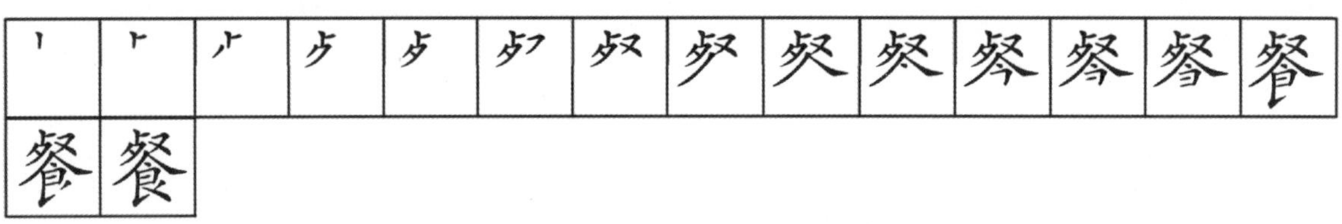

晚
wǎn

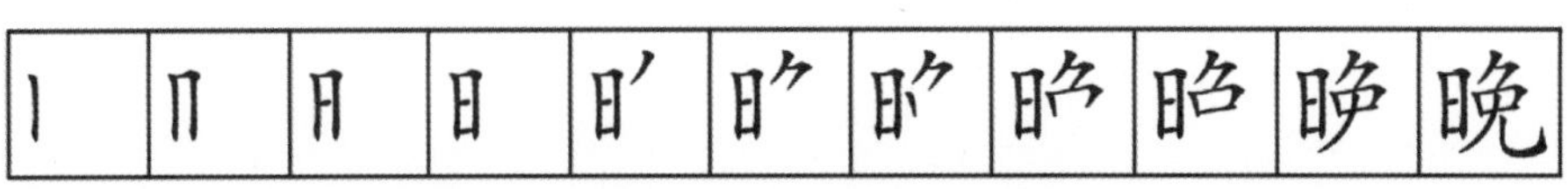

炒
chǎo

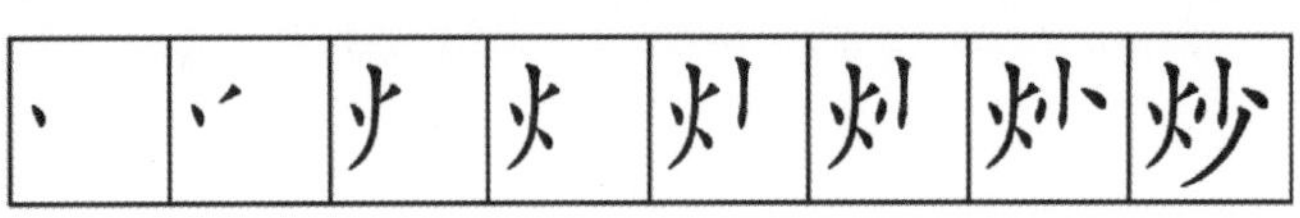

# Section 2 – Sports

Activity 10.16

Practise writing the following characters:

休

更

队

加

赛

滑

冰

次

育

早

晚

餐

炒

## Activity 10.17

Draw lines to match pairs and make logical phrases.

| | |
|---|---|
| 参加 | 更矮 |
| 跟他 | 滑得很好 |
| 比我 | 健身房 |
| 滑雪 | 比赛 |
| 走到 | 一样慢 |

## Activity 10.18

Split up the following words and write each under the correct heading according to the tone of each character. As an example, 体 has been completed for you.

体育馆　队员　晚餐　滑冰　最近

| 1st tone | 2nd tone |
|---|---|
| | |

| 3rd tone | 4th tone |
|---|---|
| 体 | |

## Activity 10.19

Draw a line to match each Chinese word with its English translation.

| | |
|---|---|
| 金牌 | everyone |
| 体操 | gymnastics |
| 次 | an occurrence |
| 分 | minute |
| 米 | gold medal |
| 炒 | to fry |
| 大家 | metre |

## Activity 10.20

Write the pinyin and tones of these sentences.

1. 她是赛车队员。
2. 他跟我一样聪明。
3. 谁拿到金牌?
4. 你最近去健身房吗?
5. 少吃炒饭对身体好。

## Activity 10.21

Below, fill in the blanks with appropriate words.

1. 我的头＿，今天不能参加比＿了。
2. 他会不会＿冰?
3. 她滑雪滑得很好，一共＿过八个＿牌。
4. ＿近我去了体＿馆两＿。
5. 午餐以后，我们去健＿房＿!

| | | | | |
|---|---|---|---|---|
| 拿 | 滑 | 育 | 最 | 赛 |
| 身 | 金 | 次 | 疼 | 吧 |

## Activity 10.22

Translate the following sentences into Chinese.

1. In what year did you come to the UK?

2. Please give everyone a look at your gold medal.

3. Although I run fast, she is still faster than me.

4. My teammate was here, but now he's gone.

5. What shall we have for dinner?

## Activity 10.23

Match the start and end of each sentence to make a logical complete sentence.

| | |
|---|---|
| 滑雪队员 | 三百米，不远。 |
| 你几点 | 去参观长城？ |
| 邮局离这儿 | 吃早餐？ |
| 她跟我 | 都滑得很好。 |
| 这是你第几次 | 一样喜欢体操。 |

## Activity 10.24

Below are some plans you made that have fallen through. Write a sentence for each to state that the activity will no longer happen as planned, and make up a reason why.

1. 跟朋友去体育馆

2. 午餐吃炒面

3. 跟队员去体育馆踢足球

4. 参观天安门广场

5. 去银行换钱

## Activity 10.25
You've received the following text from your friend. Read it and answer the questions in English.

你好！对不起，因为我感冒了，所以我不能和你参加比赛了。前天我看病的时候，医生说我应该多休息。可是因为我最近有很多作业，所以我没有时间休息。现在我的身体很不舒服。请你别生气，跟队员说一下，好吗？下个星期我会跑得更好。

1. What is your friend unable to do?
2. What did her doctor say?
3. What prevented her from following this advice?
4. What does she ask you to do for her?
5. What does she promise to do next week?

## Activity 10.26
Answer the questions for yourself in Chinese. Use complete sentences.

1. 你最喜欢看什么运动？
2. 你的城市有没有健身房？
3. 你常常参加运动比赛吗？
4. 体操、赛车、滑冰、滑雪，哪个最难？

## Activity 10.27
Translate the following passage into English.

我是体操队员。最近因为我伤了我的腿，所以我不能参加比赛。我只能去看我的体操队比赛。我最好的朋友很矮，也很瘦，她总是拿到金牌。下个星期我想和她一起去体育馆。

## Activity 10.28
Translate the following passage into Chinese.

Although I am on the ski team, I don't like competing. Every week we go to the mountains to compete. The mountains are far from my school. My teammates and I must wake up very early. Sometimes I don't have breakfast. I don't want to attend competitions anymore.

# Unit 11 Section 1 Vocabulary

| | | |
|---|---|---|
| 科目 | kēmù | (school) subject |
| 生物学 | shēngwùxué | biology |
| 化学 | huàxué | chemistry |
| 物理 | wùlǐ | physics |
| 地理 | dìlǐ | geography |
| 感兴趣 | gǎn xìngqu | to be interested in |
| 体育 | tǐyù | physical education |
| 懂 | dǒng | to understand |
| 美术 | měishù | fine arts |
| 工艺 | gōngyì | arts and crafts / design technology |
| 容易 | róngyì | easy |
| 礼堂 | lǐtáng | auditorium |
| 校长 | xiàozhǎng | head of school |
| 办公室 | bàngōngshì | office |
| 走廊 | zǒuláng | corridor |
| 食堂 | shítáng | dining hall |
| 操场 | cāochǎng | sports field |
| 经常 | jīngcháng | often |
| 很少 | hěnshǎo | rarely, few |
| 考试 | kǎoshì | test, exam |

**Online vocabulary**

| | | |
|---|---|---|
| 校服 | xiàofú | school uniform |
| 书包 | shūbāo | book bag |
| 尺子 | chǐzi | ruler |
| 橡皮 | xiàngpí | eraser |
| 同学 | tóngxué | classmate |
| 学期 | xuéqī | school term |
| 忘 | wàng | to forget |
| 带 | dài | to bring, take |
| 计算器 | jìsuànqì | calculator |

# Unit 11 Section 1 Stroke Order

**科目** kēmù

丿 二 千 禾 禾 禾 禾 禾 科

丨 冂 冃 月 目

**物理** wùlǐ

丿 𠂉 牛 牛 牜 牞 物 物

一 二 干 王 𤣩 𤣩 𤣩 𤣩 理 理 理

**化学** huàxué

丿 亻 亻 化

**学校** xuéxiào

一 十 才 木 木 木 木 木 校 校

**美术** měishù

丶 丷 䒑 䒑 羊 羊 羊 美 美

一 十 才 木 术

**工艺** gōngyì

一 十 艹 艺

**礼堂** lǐtáng

丶 ㇇ 礻 礻 礼

丨 丨 ⺌ ⺌ 尚 尚 尚 尚 堂 堂 堂

**办公室** bàngōngshì

𠃌 力 办 办

丿 八 公 公

丶 丶 宀 宀 宀 宀 宀 室 室

**食堂** shítáng

丿 人 亼 今 今 今 食 食 食

# 教育 Education

## Section 1 – School and subjects

Activity 11.1

Practise writing the following characters:

科

目

物

理

化

校

美

术

艺

礼

堂

办

公

室

食

### Activity 11.2
Draw lines to match each activity with a logical corresponding location.

| | |
|---|---|
| 上体育课 | 礼堂 |
| 吃午饭 | 办公室 |
| 参加表演 | 图书馆 |
| 看校长 | 操场 |
| 做作业 | 食堂 |

### Activity 11.3
Split up the following words and write each under the correct heading according to the tone of each character. As an example, 物 has been completed for you.

物理　办公室　科目　学期　体育

| 1st tone | 2nd tone |
|---|---|
| | |
| **3rd tone** | **4th tone** |
| | 物 |

### Activity 11.4
Draw a line to match each Chinese word with its English translation.

| | |
|---|---|
| 化学 | chemistry |
| 生物 | physics |
| 物理 | history |
| 科学 | maths |
| 地理 | physical education |
| 历史 | geography |
| 数学 | biology |
| 体育 | arts and crafts |
| 工艺 | science |

### Activity 11.5
Write the pinyin and tones of these sentences.

1. 我觉得工艺很不容易。
2. 我忘了带我的尺子。
3. 我们经常有中文考试。
4. 你的同学和你一样喜欢美术课吗？
5. 这条走廊长得不得了！

### Activity 11.6
Below, fill in the blanks with appropriate words.

1. 我觉得生＿学比＿学更有意思。
2. 学＿离我家不＿，走路两分钟就到了。
3. 我的地＿老师说得很＿，我＿不懂。
4. 你的同＿都在哪儿？他们在＿廊里。
5. 我下午在操＿参加比赛。

| | | | | |
|---|---|---|---|---|
| 化 | 听 | 快 | 场 | 校 |
| 学 | 远 | 走 | 物 | 理 |

## Activity 11.7

Translate the following sentences into Chinese.

1. My favourite subjects are physics and chemistry.
2. I waited until 5pm, then I left.
3. Your paintings are beautiful; you should be a fine artist!
4. I saw him in the dining hall already.
5. Please go to the head's office after biology class.

## Activity 11.8

Match the start and end of each sentence to make a logical complete sentence.

| | |
|---|---|
| 化学作业不容易， | 不饿了。 |
| 他游泳游得很快， | 你看懂了没有？ |
| 我吃好了， | 睡到了十一点！ |
| 新鞋 | 你买到了吗？ |
| 我昨天很累， | 经常拿到金牌。 |

## Activity 11.9

Below, rewrite the sentences, putting each adverb in the correct place.

1. 我在办公室看到他。 （经常）
2. 他的课本是蓝色的。 （也）

## Activity 11.9 (continued)

3. 超市很早就关门了。 （每天）
4. 我朋友考试考得很好。 （总是）
5. 我骑自行车上学。 （很少）

## Activity 11.10

Read the passage about Tiffany's school day and answer the questions in English.

我很少努力学习，因为我觉得生物学不难，只要好好听老师的话就可以。我也总是考得很好，所以我很少做作业。前天我们考了生物考试，可是我忘了老师教的内容！下课以后，我和同学去食堂吃饭了，可是我不想吃饭，也不想和大家聊天。我太伤心了。

1. In what subject did Tiffany have an exam?
2. How does she normally feel about that subject?
3. What does she normally do to learn the material?
4. What happened the day of the exam?
5. Where did she go after class?

## Activity 11.11

Answer the questions for yourself in Chinese. Use complete sentences.

1. 你对哪个科目最感兴趣?

2. 你觉得哪个科目最难?

3. 你的学校离家近不近? 上学要多长时间?

4. 你喜欢穿校服吗?

5. 你的中文班一共有多少个同学?

## Activity 11.12

Translate the following passage into English.

我的校服难看极了! 大家都要穿更好看的衣服。可是校长说了，同学们都要穿校服。只有上体育课时，可以穿自己的运动服。我去过美国，美国学生不用穿校服。我觉得在美国上学多好。

## Activity 11.13

Translate the following passage into Chinese.

I am really interested in fine arts. My older sister works in an art museum. The entrance tickets are free, so I go to visit frequently. I bring my notebook, too. My sister has seen a famous artist there on two occasions.

# Unit 11 Section 2 Vocabulary

| | | |
|---|---|---|
| 严格 | yángé | strict |
| 汉字 | hànzì | Chinese character |
| 问题 | wèntí | question |
| 习惯 | xíguàn | habit, to be used to |
| 完 | wán | to finish |
| 错 | cuò | mistake, wrong |
| 问 | wèn | to ask |
| 糟糕 | zāogāo | terrible |
| 有空 | yǒukòng | to have free time |
| 父亲 | fùqīn | father |
| 母亲 | mǔqīn | mother |
| 高考 | gāokǎo | college entrance exam |
| 第 | dì | (prefix indicating ordinal number) |
| 教育 | jiàoyù | education |
| 设备 | shèbèi | equipment |

**Online vocabulary**

| | | |
|---|---|---|
| 实习 | shíxí | to intern, internship, work experience |
| 经济 | jīngjì | economy |

# Unit 11 Section 2 Stroke Order

懂
dǒng

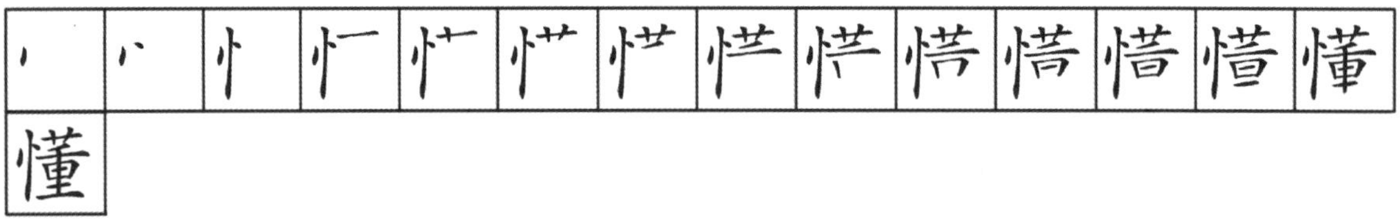

感兴趣
gǎn xìngqù

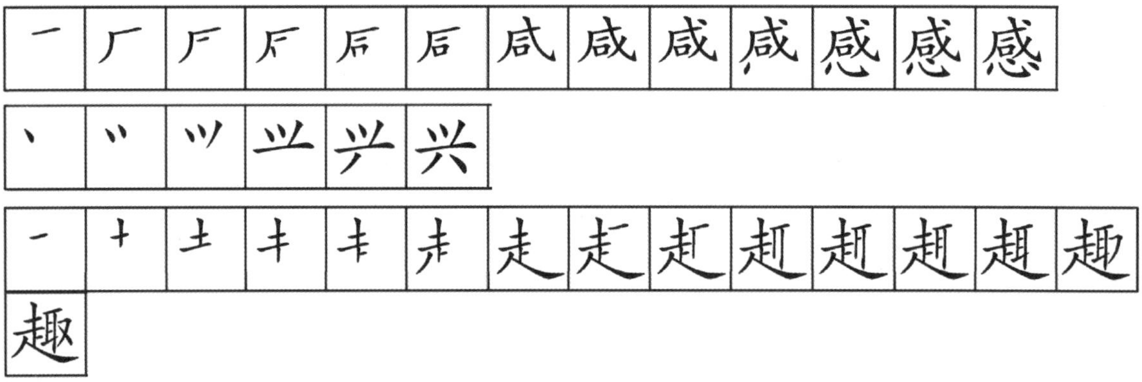

问题
wèntí

完
wán

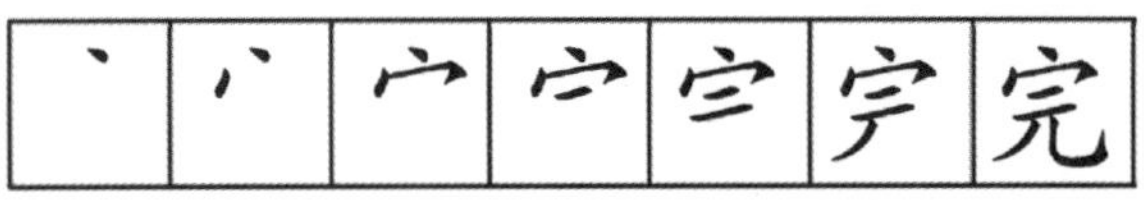

错
cuò

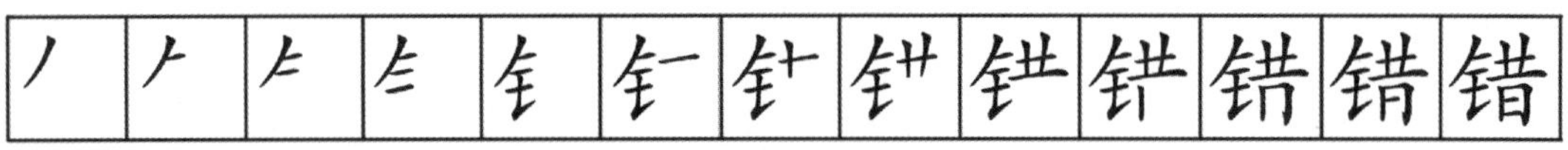

有空
yǒukòng

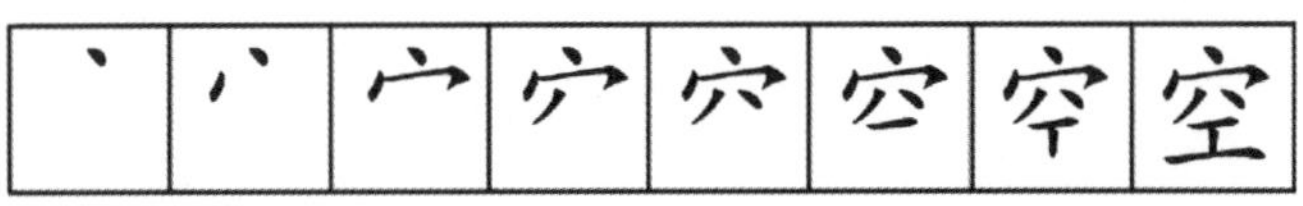

父亲
fùqīn

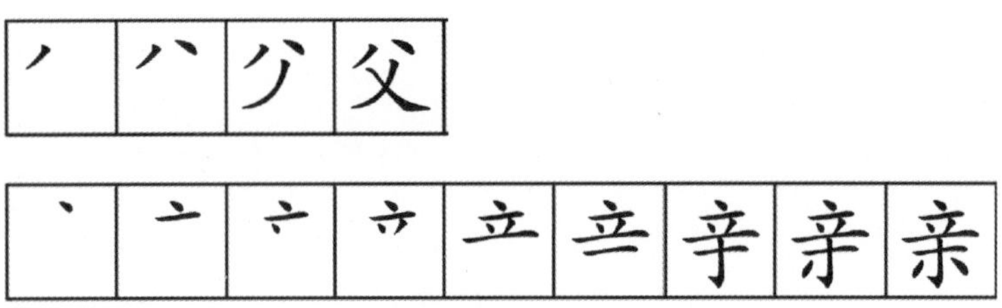

母亲
mǔqīn

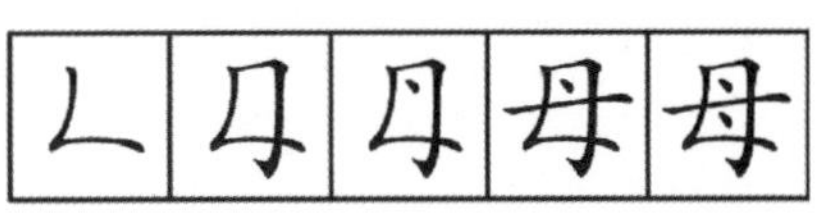

高考
gāokǎo

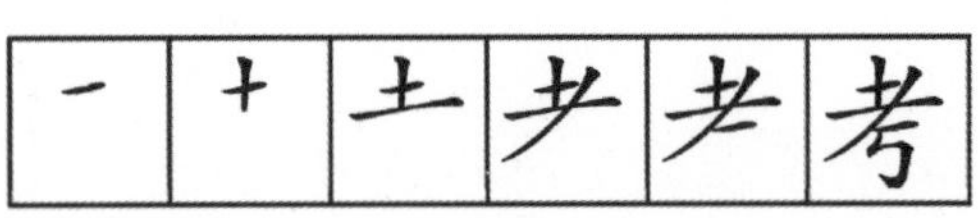

第
dì

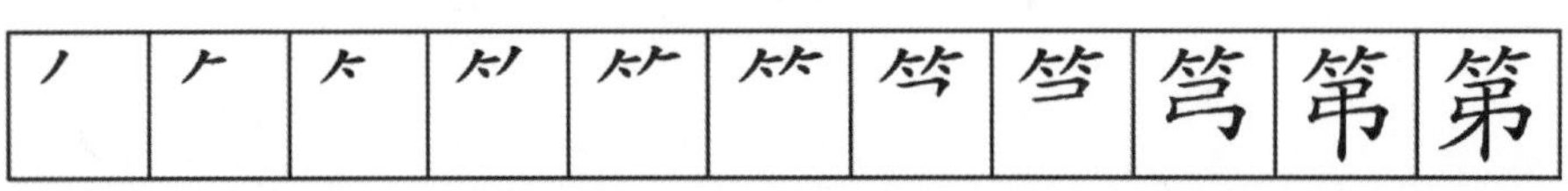

教育
jiàoyù

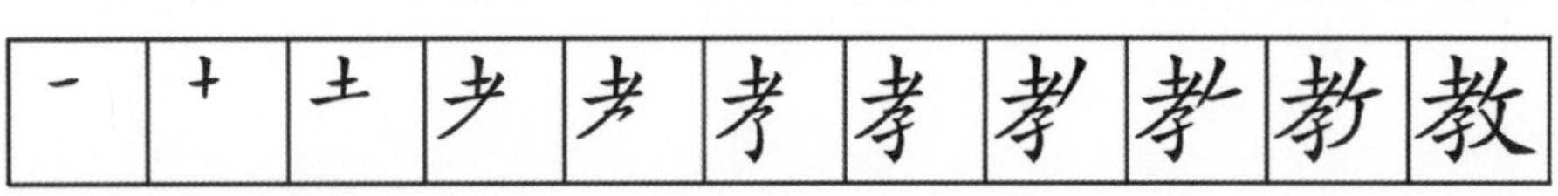

# Section 2 – Plans for the future

Activity 11.16

Practise writing the following characters:

| | | | | | | | | | | | |
|---|---|---|---|---|---|---|---|---|---|---|---|
| 懂 | | | | | | | | | | | |
| 感 | | | | | | | | | | | |
| 兴 | | | | | | | | | | | |
| 趣 | | | | | | | | | | | |
| 题 | | | | | | | | | | | |
| 完 | | | | | | | | | | | |
| 错 | | | | | | | | | | | |
| 空 | | | | | | | | | | | |
| 父 | | | | | | | | | | | |
| 母 | | | | | | | | | | | |
| 亲 | | | | | | | | | | | |
| 考 | | | | | | | | | | | |
| 第 | | | | | | | | | | | |
| 教 | | | | | | | | | | | |

### Activity 11.17
Draw lines to match pairs and make logical phrases.

| | |
|---|---|
| 第 | 高考 |
| 问 | 三次 |
| 考 | 实习 |
| 严格的 | 有空 |
| 在博物馆 | 问题 |
| 周末 | 母亲 |

### Activity 11.18
Split up the following words and write each under the correct heading according to the tone of each character. As an example, 高 has been completed for you.

高考　教育　感兴趣　有空　父亲

| 1st tone | 2nd tone |
|---|---|
| 高 | |
| **3rd tone** | **4th tone** |
| | |

### Activity 11.19
Draw a line to match each Chinese word with its English translation.

| | |
|---|---|
| 习惯 | economy |
| 经济 | terrible |
| 设备 | habit, to be used to |
| 实习 | equipment |
| 糟糕 | to intern, internship |

### Activity 11.20
Write the pinyin and tones of these sentences.

1. 父亲严格得不得了！

2. 健身房的设备都很新。

### Activity 11.20 (continued)

3. 经济学很不容易！

4. 我在办公室等你。你考完试，请到那儿去跟我见面。

5. 我生病了，不能参加表演了。糟糕！

### Activity 11.21
Below, fill in the blanks with appropriate words.

1. 我还没有看☐比赛，可是我已经☐☐谁会拿☐金牌。

2. 这次☐试，你别写☐了！

3. 父亲经常对我说，☐育不是免费的。

4. 老师的问☐你都听☐了吗？

5. 我对在银行实☐没有兴趣。

| | | | | |
|---|---|---|---|---|
| 错 | 懂 | 考 | 习 | 完 |
| 题 | 教 | 到 | 知道 | |

## Activity 11.22

Translate the following sentences into Chinese.

1. I wrote this Chinese character incorrectly!
2. I am interested in geography. I am not interested in design technology.
3. This is my fourth time visiting Xi'an.
4. My mother hasn't had any free time recently.
5. He already finished sitting the college entrance exam.

## Activity 11.23

Match the start and end of each sentence to make a logical complete sentence.

| | |
|---|---|
| 我的同学学习很努力， | 她是大学校的校长。 |
| 老师不严格， | 可是她只会在家里睡觉。 |
| 我叔叔很酷， | 我每次看到他，他都戴新帽子。 |
| 奶奶每天有空， | 她总是问老师很多问题。 |
| 阿姨很聪明， | 她不给我们很多作业。 |

## Activity 11.24

Below are some activities and time durations. Create two sentences for each prompt: the first sentence should indicate that the activity was done for a time and then stopped; the second should indicate that the activity is continuing now.

1. 爸爸当护士（两年）
2. 我弟弟玩手机（二十分钟）

Activity 11.24 (continued)

3. 他的同学唱歌（一个半小时）

## Activity 11.25

Read the passage about Xiao Peng taking the college entrance exam and answer the questions in English.

我去年上高三。我六月考了高考，考得不太好。我对自己很严格，觉得很不高兴。现在我不上学了，每天去图书馆自己学习。我还有一些同学，他们跟我一样考得不好，有时候我跟他们一起学习。我们都要再考一次。

1. When did Xiao Peng take the college entrance exam?
2. How does Xiao Peng describe himself?
3. How did he feel about his results?
4. What does Xiao Peng do every day?
5. What has Xiao Peng decided to do this year?

## Activity 11.26

Answer the questions for yourself in Chinese. Use complete sentences.

1. 你做完作业以后，喜欢做什么？
2. 你常常写错什么汉字？
3. 这个学期的第一天，你觉得怎么样？
4. 你做过实习吗？在哪儿？实习怎么样？

### Activity 11.27
Translate the following passage into English.

中国的高考每年六月七日举办。高考一共三天。高考很难，中国学生都觉得压力很大。高中生学习很多科目：语文、数学和英语是最重要的。考完了高考，学生要等几个星期才知道考得怎么样。

| | |
|---|---|
| 举办 | jǔbàn = to hold an event |
| 压力 | yālì = pressure |
| 重要 | zhòngyào = important |
| 才 | cái = "only then" |

### Activity 11.28
Translate the following passage into Chinese.

I can't forget the first time I ate Chinese noodles. I was eating lunch in the dining hall with my classmates. The noodles were so delicious! My friend said to me, 'Eat more!' I ate too much and got a headache. Next time I will eat more slowly.

# Unit 12 Section 1 Vocabulary

| | | |
|---|---|---|
| 迷路 | mí lù | to be lost |
| 过马路 | guò mǎlù | to cross the road |
| 往 | wǎng | to go towards |
| 看见 | kànjiàn | to see |
| 红绿灯 | hónglǜdēng | traffic light |
| 十字路口 | shízì lùkǒu | intersection |
| 拐 | guǎi | to turn (a corner) |
| 一直 | yīzhí | straight, continuously |
| 地图 | dìtú | map |
| 小镇 | xiǎo zhèn | small town |
| 汽车站 | qìchē zhàn | bus stop |
| 安静 | ānjìng | quiet, peaceful |
| 环境 | huánjìng | environment |
| 郊区 | jiāoqū | suburbs |
| 书店 | shū diàn | book store |
| 生活 | shēnghuó | life, to live |
| 地点 | dìdiǎn | site, location |
| 交通 | jiāotōng | traffic |
| 快 | kuài | fast, quickly |
| 那么 | nàme | like that, so much |

**Online vocabulary**

| | | |
|---|---|---|
| 湖 | hú | lake |
| 街 | jiē | street |
| 散步 | sàn bù | to take a walk |
| 窄 | zhǎi | narrow |
| 开车 | kāi chē | to drive |

# Unit 12 Section 1 Stroke Order

| Word | Pinyin | Stroke order shown for |
|---|---|---|
| 迷路 | mí lù | 迷 |
| 往 | wǎng | 往 |
| 住 | zhù | 住 |
| 红绿灯 | hónglǜdēng | 灯 |
| 一直 | yīzhí | 直 |
| 小镇 | xiǎo zhèn | 镇 |
| 安静 | ānjìng | 静 |
| 环境 | huánjìng | 环, 境 |
| 生活 | shēnghuó | 活 |
| 交通 | jiāotōng | 交, 通 |
| 街 | jiē | 街 |

# 我的家乡 My Hometown

## Section 1 – In my town

Activity 12.1

Practise writing the following characters:

迷

往

住

灯

直

镇

静

环

境

活

交

通

街

### Activity 12.2
Draw lines to match pairs and make logical phrases.

| | |
|---|---|
| 一直 | 目的地 |
| 往左 | 车 |
| 过 | 拐 |
| 上 | 往前走 |
| 到达 | 马路 |

### Activity 12.3
Split up the following words and write each under the correct heading according to the tone of each character. As an example, 小 has been completed for you.

小镇　地点　迷路　安静　交通

| 1st tone | 2nd tone |
|---|---|
| | |

| 3rd tone | 4th tone |
|---|---|
| 小 | |

### Activity 12.4
Draw a line to match each Chinese word with its English translation.

| | |
|---|---|
| 散步 | narrow |
| 窄 | map |
| 郊区 | to turn |
| 地图 | to take a walk |
| 湖 | lake |
| 小镇 | suburbs |
| 拐 | small town |

### Activity 12.5
Write the pinyin and tones of these sentences.

1. 你习惯了这儿的生活吗?

2. 火车站离这儿太远了。

3. 我最近没看到他。下个星期我们会去公园散步。

4. 我忘了，到书店要左拐还是右拐?

5. 城市的交通问题比郊区更严重。

### Activity 12.6
Below, fill in the blanks with appropriate words.

1. 我的生____有意思极了!

2. ____书馆里的书跟书____一样多。

3. 他____在郊区，离公司不近，所以每天都开车去上____。

4. 这是小____，没有很多小街，你不会____路。

5. 那边的红________你看见吗? 到那儿以后，就要____右拐。

| | | | | |
|---|---|---|---|---|
| 住 | 往 | 迷 | 活 | 店 |
| 镇 | 班 | 图 | 绿灯 | |

## Activity 12.7

Translate the following sentences into Chinese.

1. This little town is extremely quiet.
2. Don't drive so fast!
3. Does anyone have a map? I'm lost.
4. When you get to the crossroads, go straight ahead.
5. Next time I won't drink so much fizzy drink.

## Activity 12.8

Match the start and end of each sentence to make a logical complete sentence.

| | |
|---|---|
| 这个城市 | 我四点就回到家了。 |
| 这条街太窄了， | 很感兴趣。 |
| 因为交通很好， | 有个很大的湖。 |
| 汽车已经到了车站， | 车不能开进去！ |
| 生物老师对动物 | 快过马路吧！ |

## Activity 12.9

Add 就 into each sentence below, and then write the English translation.

1. 作业很难，我问老师怎么办。

## Activity 12.9 (continued)

2. 我八点半到学校。
3. 我不知道他的名字。
4. 你努力学习会考得好。
5. 汽车站在我家对面。

## Activity 12.10

Read the passage about learning to drive and answer the questions in English.

我是加拿大人。在我的国家，十四岁的年轻人可以学开车。我觉得开车不难：第一，我的小镇没有很多小路，我不会迷路；第二，路不窄，十字路口也很少；第三，大家开得很慢。如果我开得不好，很少会有人生气，大家知道我在学习开车。

1. Which nationality is the speaker?
2. Where he lives, what is the minimum driving age?
3. The speaker thinks driving is easy; what three reasons does he give?
4. What would happen if the speaker makes a mistake while driving?

## Activity 12.11

Answer the questions for yourself in Chinese. Use complete sentences.

重要 zhòngyào = important

1. 在你的生活中，什么事儿对你最重要？
2. 你带什么东西去旅游？
3. 请说说你的小镇怎么样。

## Activity 12.12

Translate the following passage into English.

明明：体育馆怎么走？
叔叔：过马路。一直往前走。你到了十字路口，往左拐。再过两个红绿灯。到了第三个十字路口，就往右拐。往前走三分钟就到了。对面是中国银行。

## Activity 12.13

Translate the following passage into Chinese.

I live in a small town. Life here is very quiet and there aren't many shops. Recently I went to the city to buy a new mobile phone. I waited for 10 minutes at the bus stop. Although the city is far away, the traffic was good, so I arrived just one hour later. That fast! I was so happy.

# Unit 12 Section 2 Vocabulary

| | | |
|---|---|---|
| 雾 | wù | fog, mist |
| 天气预报 | tiānqì yùbào | weather forecast |
| 小雨 | xiǎoyǔ | light rain |
| 暖和 | nuǎnhuo | warm |
| 晴天 | qíngtiān | fine day |
| 阴天 | yīntiān | cloudy day |
| 凉快 | liángkuai | cool |
| 离开 | líkāi | to leave |
| 就 | jiù | "just", "as soon as" |
| 桥 | qiáo | bridge |
| 早 | zǎo | early |
| 才 | cái | "only", "not until" |
| 祝 | zhù | to wish (somebody well) |

**Online vocabulary**

| | | |
|---|---|---|
| 毒品 | dúpǐn | drugs |
| 小偷 | xiǎo tōu | thief |
| 穷 | qióng | poor |
| 乞丐 | qǐgài | beggar |
| 住房 | zhùfáng | housing |
| 富有 | fùyǒu | rich |
| 脏 | zāng | dirty |

# Unit 12 Section 2 Stroke Order

同学
tóngxué

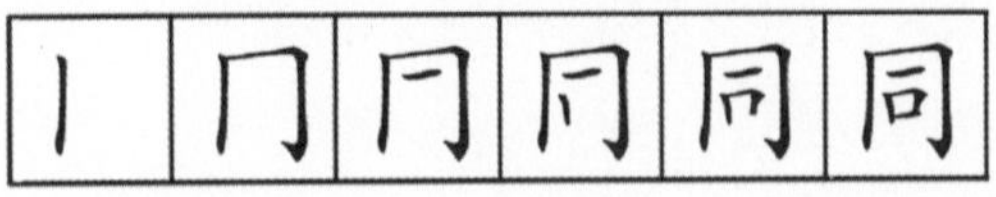

带
dài

容易
róngyì

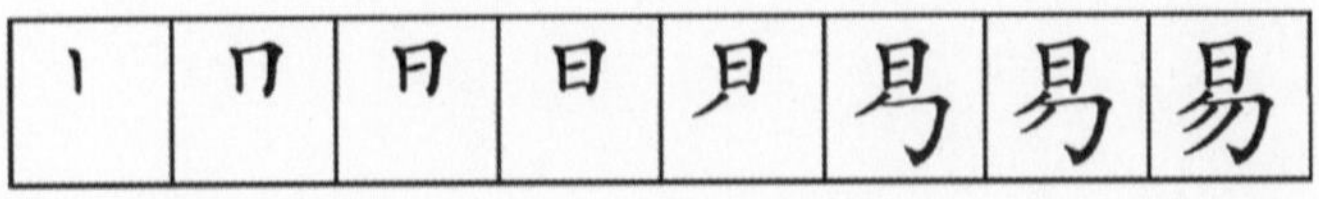

天气预报
tiānqì yùbào

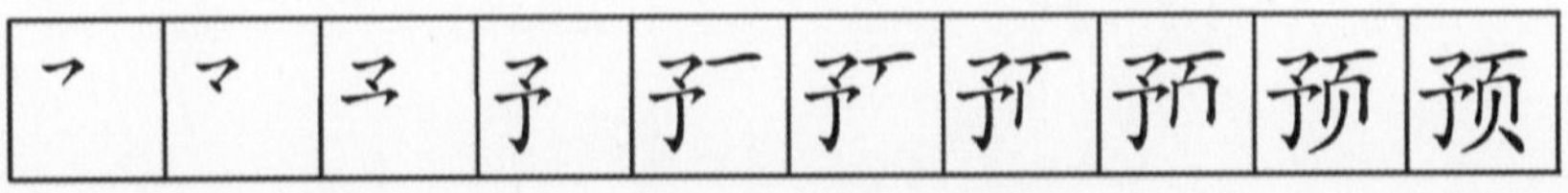

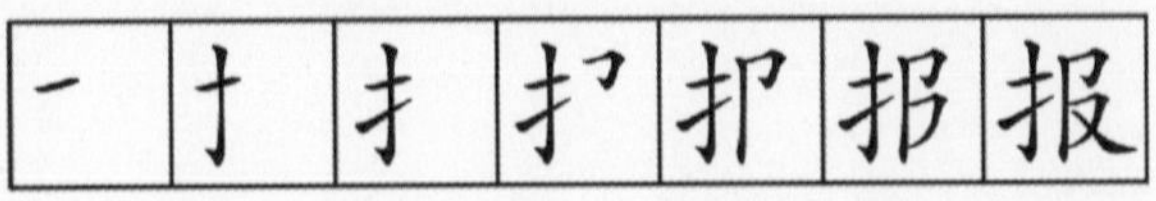

暖和
nuǎnhuo

凉快
liángkuai

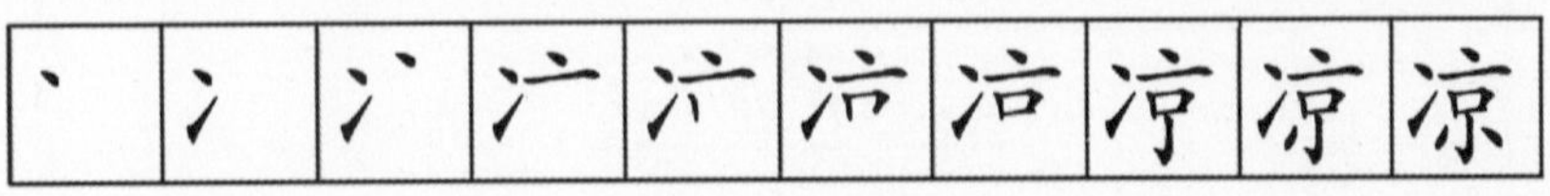

晴天
qíngtiān

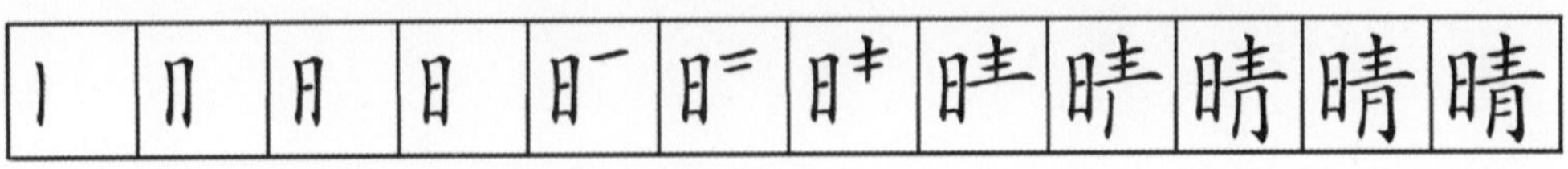

阴天
yīntiān

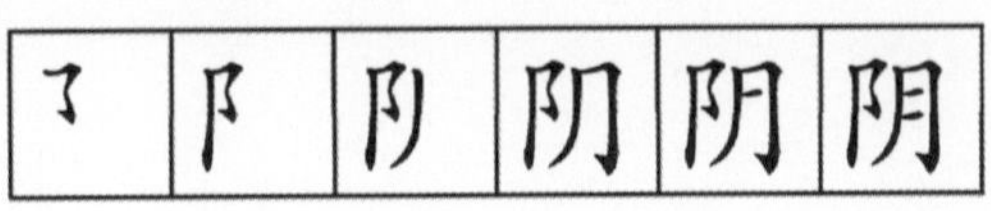

就
jiù

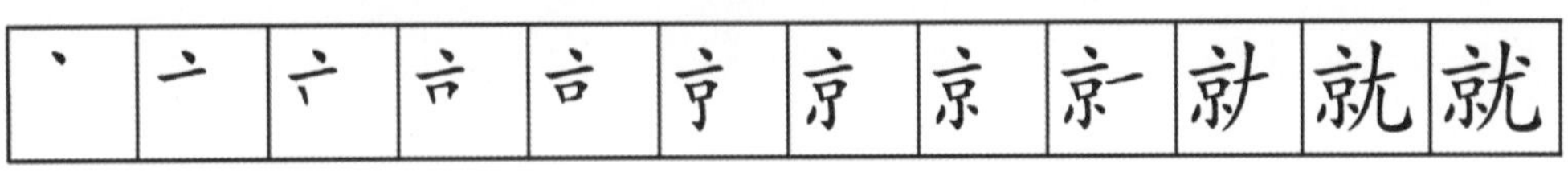

才
cái

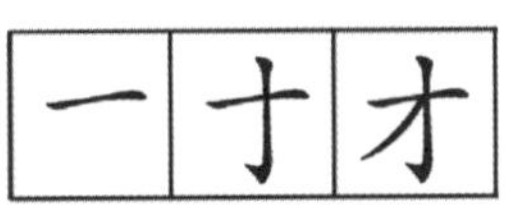

桥
qiáo

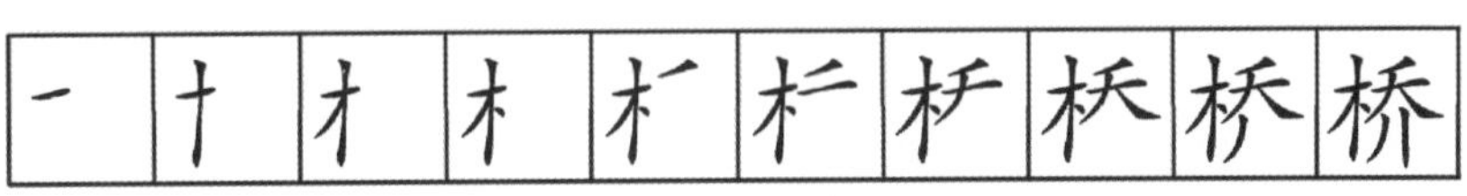

祝
zhù

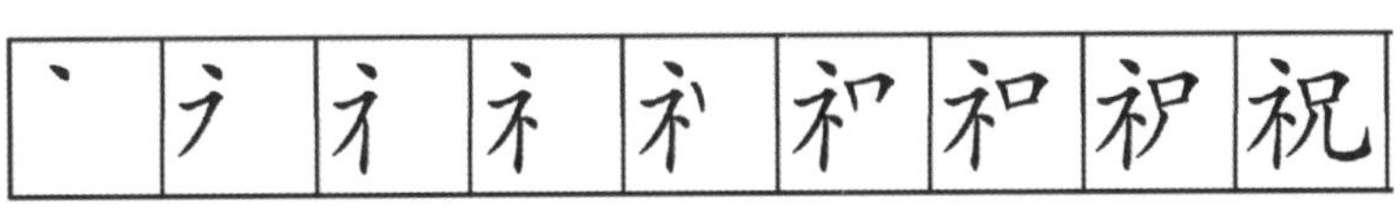

# Section 2 – Life in my local area

Activity 12.14
Practise writing the following characters:

同
带
容
易
预
报
暖
凉
晴
阴
就
才
桥
祝

### Activity 12.15

Draw lines to match the words and their simple definitions.

| | |
|---|---|
| 乞丐 | 拿走东西的人 |
| 富有 | 没有钱 |
| 小偷 | 要钱的人 |
| 毒品 | 有很多钱 |
| 穷 | 对身体不好的东西 |

### Activity 12.16

Split up the following words and write each under the correct heading according to the tone of each character. As an example, 小 has been completed for you.

小偷　　容易　　预报　　晴天　　离开

| 1st tone | 2nd tone |
|---|---|
| | |
| 3rd tone | 4th tone |
| 小 | |

### Activity 12.17

Draw a line to match each Chinese word with its English translation.

| | |
|---|---|
| 刮风 | cool |
| 有雾 | warm |
| 暖和 | misty |
| 下雪 | sunny day |
| 晴天 | windy |
| 凉快 | snowing |
| 阴天 | cloudy day |

### Activity 12.18

Write the pinyin and tones of these sentences.

1. 伦敦现在没有以前那么脏了！
2. 请别离开我！
3. 他很富有，一看新车就买。
4. 在这里，穷人不少，可是乞丐不多。
5. 最近天气很暖和。

### Activity 12.19

Below, fill in the blanks with appropriate words.

1. 我的____学都很富有。
2. 天气________说明天会更____快。
3. 今天是____天，你____足球来海滩玩儿吧！
4. 看懂地____不太____，我经常____路。
5. 上午交通不好，我十点____到了办公____。

| | | | | | |
|---|---|---|---|---|---|
| 才 | 室 | 同 | 图 | 凉 | |
| 迷 | 带 | 晴 | 容易 | 预报 | |

## Activity 12.20

Translate the following sentences into Chinese.

1. I wish you good health!

2. Today is overcast.

3. The bookshop is just there; cross the bridge and you'll see it.

4. The weather is sometimes warm, sometimes cool.

5. Your little brother has only four Yuan. Please bring extra cash for him.

## Activity 12.21

Match the start and end of each sentence to make a logical complete sentence.

| | |
|---|---|
| 有人说毒品问题很严重， | 才知道她是那么好的朋友。 |
| 我的手很脏， | 可是我从来没看到过什么毒品。 |
| 他以前有电脑， | 我去洗一洗。 |
| 你离开她， | 旅途愉快！ |
| 祝你 | 可是被小偷偷走了。 |

被 bèi = (indicates passive)
偷 tōu = to steal

## Activity 12.22

Read the passage about Danielle's living situation and answer the questions in English.

我的父母不穷，可是他们也不太富有。因为我的城市有住房问题，郊区比城市更贵，所以我们住在市中心。我们的家很小，家外边也很脏。我想当大公司的经理，这样才能赚钱，住在更好的地方。

1. How does Danielle describe her family's financial situation?

2. Where do they live, and why?

3. What does Danielle describe as being dirty?

4. What are her plans for the future?

### Activity 12.23

Answer the questions for yourself in Chinese. Use complete sentences.

1. 小偷偷过你的东西吗？

2. 你看到路上的乞丐会做什么？

3. 出去以前，你会不会看天气预报？

4. 你出门玩儿的时候，会带什么东西？

5. 你为什么喜欢你的小镇？

| | |
|---|---|
| 偷 | tōu = to steal |
| 的时候 | de shíhòu = when |

### Activity 12.24

Translate the following passage into English.

我的小镇很好玩。虽然很小，可是有很多好地方可以去。这儿有三个博物馆，两个体育馆、一个电影院和两个公园，都是免费的。还有一条古街，那儿有不少的商店和饭馆。

| | |
|---|---|
| 找 | zhǎo = to find |

### Activity 12.25

Translate the following passage into Chinese.

Looking at maps is my hobby. Every week I look at maps with my friend. Sometimes I buy maps at the bookstore; sometimes friends buy them for me. I have more than five hundred at home. My town's library also has many maps.

# Unit 13 Section 1 Vocabulary

| | | |
|---|---|---|
| 节日 | jiérì | holiday, festival |
| 春节 | Chūnjié | Spring Festival |
| 灯笼 | dēnglóng | lantern |
| 庆祝 | qìngzhù | to celebrate |
| 圣诞节 | Shèngdànjié | Christmas |
| 礼物 | lǐwù | gift |
| 中秋节 | Zhōngqiūjié | Mid-Autumn Festival |
| …的时候 | de shíhòu | when... |
| 月饼 | yuèbǐng | moon cake |
| 月亮 | yuèliang | the moon |
| 端午节 | Duānwǔjié | Dragon Boat Festival |
| 龙舟 | lóngzhōu | dragon boat |
| 粽子 | zòngzi | glutinous rice wrapped in leaves and boiled |
| 复活节 | Fùhuójié | Easter |
| 中国新年 | Zhōngguó Xīnnián | Chinese New Year |
| 重要 | zhòngyào | important |
| 过年 | guònián | to celebrate new year |
| 或者 | huòzhě | or |
| 新年快乐 | Xīnnián kuàilè | Happy new year! |
| 红包 | hóngbāo | red envelope filled with money |
| 舞龙 | wǔlóng | dragon dance |
| 舞狮 | wǔshī | lion dance |
| 鞭炮 | biànpào | firecrackers |
| 属 | shǔ | to belong to, be affiliated with |

**Online vocabulary**

| | | |
|---|---|---|
| 季节 | jìjié | season |
| 赛马 | sàimǎ | horse racing |
| 春天 | chūntiān | spring |
| 夏天 | xiàtiān | summer |
| 秋天 | qiūtiān | autumn |
| 冬天 | dōngtiān | winter |

# Unit 13 Section 1 Stroke Order

重要
zhòngyào

月饼
yuèbǐng

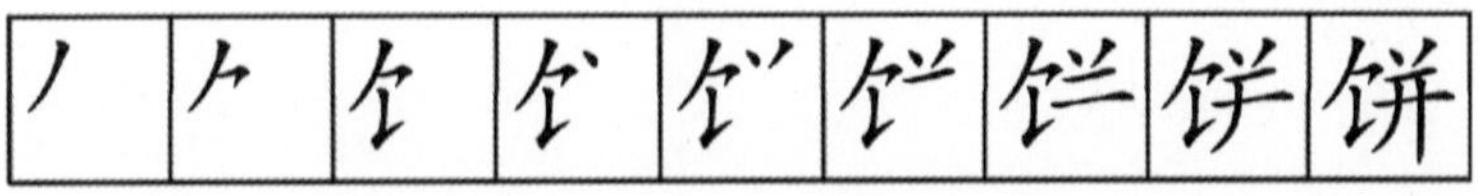

端午节
duānwǔjié

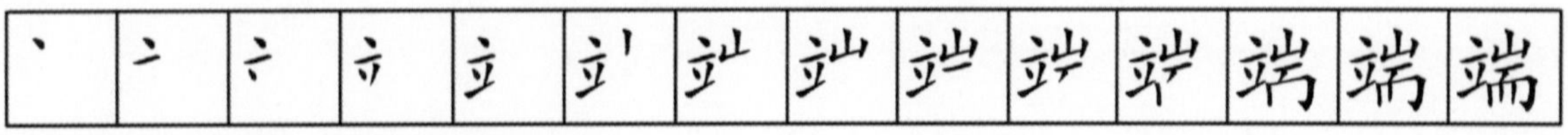

龙舟
lóngzhōu

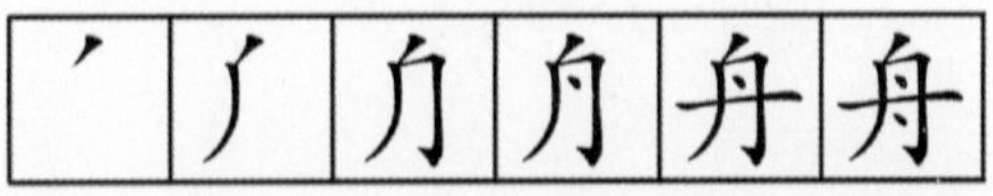

复活节
fùhuójié

或
huò

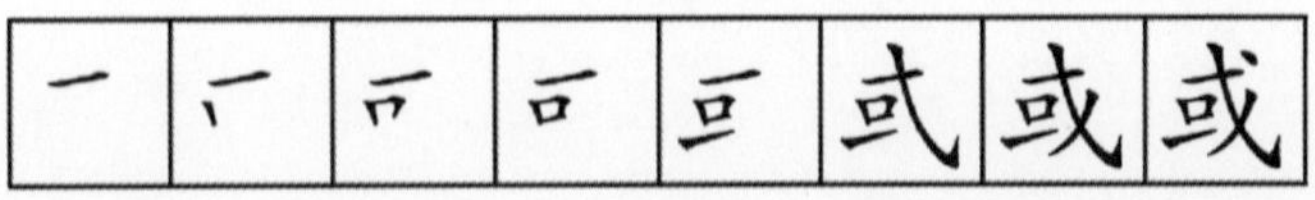

舞龙
wǔlóng

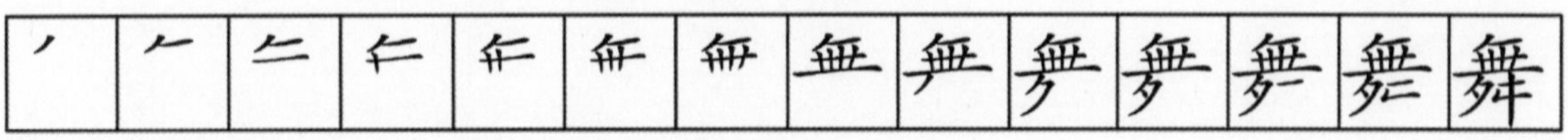

舞狮
wǔshī

季节
jìjié

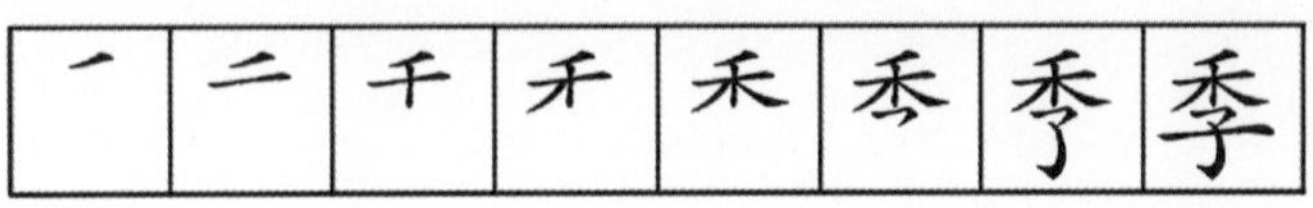

春天
chūntiān

夏天
xiàtiān

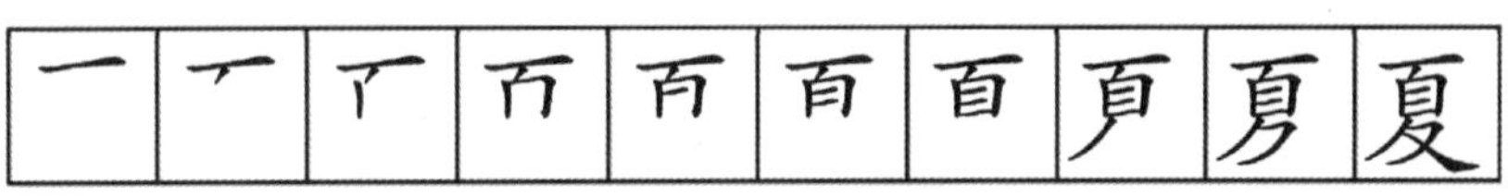

秋天
qiūtiān

冬天
dōngtiān

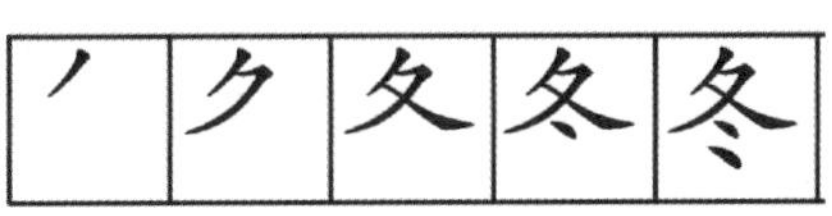

# 中国传统 Chinese Traditions
## Section 1 – Festivals

Activity 13.1
Practise writing the following characters:

重
饼
端
龙
舟
复
或
舞
狮
季
春
夏
秋
冬

## Activity 13.2

Draw lines to match the holiday with the month(s) in the Western calendar in which it occurs.

| | |
|---|---|
| 端午节 | 十二月 |
| 圣诞节 | 九月、十月 |
| 中秋节 | 五月、六月 |
| 复活节 | 一月、二月 |
| 春节 | 三月、四月、五月 |

## Activity 13.3

Split up the following words and write each under the correct heading according to the tone of each character. As an example, 月 has been completed for you.

月饼　舞狮　秋天　礼物　春节

| 1st tone | 2nd tone |
|---|---|
| | |

| 3rd tone | 4th tone |
|---|---|
| | 月 |

## Activity 13.4

Draw a line to match each Chinese word with its English translation.

| | |
|---|---|
| 庆祝 | firecrackers |
| 灯笼 | dragon boat |
| 粽子 | lantern |
| 鞭炮 | dragon dance |
| 月亮 | the moon |
| 龙舟 | to celebrate |
| 舞龙 | rice packet |

## Activity 13.5

Write the pinyin and tones of these sentences.

1. 我家人不庆祝圣诞节。

2. 她每年赛龙舟，可是没拿到过金牌。

3. 端午节对你重要吗？

4. 我想住在郊区或者小镇。

5. 过年的时候，我出门散步，看到了小孩子放鞭炮。

## Activity 13.6

Below, fill in the blanks with appropriate words.

1. 明天是爷爷的生日。你要给他送什么☐☐？

2. 中☐节的时候，我的☐学送给我月☐吃。

3. 去年过☐☐☐的时候，我去了德国旅游。

4. 学校的☐☐活动非常好玩。

秋　送　同　饼　新
礼物　春节　复活节

## Activity 13.7

Translate the following sentences into Chinese.

1. The Mid-Autumn Festival is very important to me.
2. Dragon boat racing is not easy!
3. Which is your favourite holiday?
4. The day after tomorrow, I'll give him a red envelope.
5. The summertime is okay, but I like winter even more.

## Activity 13.8

Match the start and end of each sentence to make a logical complete sentence.

| | |
|---|---|
| 你买到 | 庆祝过圣诞节没有? |
| 你有没有 | 月饼了没有? |
| 昨天你吃 | 参加过龙舟比赛? |
| 你昨天放 | 红衣服了没有? |
| 你 | 鞭炮了没有? |

## Activity 13.9

Your friend participates in the following activities and you want to ask her how long she has been doing each. For each item, write the question you would ask her.

1. 学习法语
2. 弹钢琴

## Activity 13.9 (continued)

3. 跳舞
4. 赛马
5. 当足球队员

## Activity 13.10

Read the passage about what Anastasia does in various seasons and answer the questions in English.

我什么季节都喜欢。夏天最好，我放三个月的假。虽然我晚上在饭馆工作，可是还没有学期中那么忙。冬天我喜欢庆祝新年，这时候我会考虑明年要做什么。秋天天气很凉快，很舒服！我住在俄罗斯，这儿的风景很美。最后，春天我常常带着一本书去海滩休息，海边很安静！

1. What is Anastasia's primary occupation?
2. What does she like best in the winter?
3. Where does she live?
4. What does she like about the autumn?
5. What does she like to do in the spring?

## Activity 13.11
Answer the questions for yourself in Chinese. Use complete sentences.

1. 你最喜欢的节日是什么？为什么？

2. 你最喜欢的季节是什么？为什么？

3. 你有没有吃过月饼？粽子呢？

4. 你有空的时候喜欢做什么？

5. 你学汉语已经有多长时间了？

## Activity 13.12
Translate the following passage into English.

我的生日是二月五日。那时候是冬天，天气很冷，我喜欢去外边滑冰。因为我母亲说滑雪太贵了，所以我们很少去。今年她给我送了滑雪场的门票。我高兴极了，这是最好的生日礼物！

## Activity 13.13
Translate the following passage into Chinese.

Chinese New Year is very important to my family. When we celebrate the new year, we make lots of dumplings together. On the street there is a dragon dance; we all go out to watch. Last year was very cold and a big snow fell! This year the forecast is better.

# Unit 13 Section 2 Vocabulary

| | | |
|---|---|---|
| 邀请 | yāoqǐng | to invite |
| 感谢 | gǎnxiè | "Thank you!" |
| 干杯 | gānbēi | "Cheers!" |
| 交朋友 | jiāo péngyou | to make friends |
| 以为 | yǐwéi | to mistakenly believe |
| 明白 | míngbai | to understand |
| 又……又…… | yòu...yòu... | both…and… |
| 您贵姓 | Nín guìxìng? | "What's your name?" (polite) |
| 位 | wèi | measure word for people (conveys respect) |

**Online vocabulary**

| | | |
|---|---|---|
| 文化 | wénhuà | culture |
| 礼貌 | lǐmào | polite |
| 早上好 | Zǎoshàng hǎo! | "Good morning!" |
| 晚上好 | Wǎnshàng hǎo! | "Good evening!" |
| 客人 | kèrén | guest |
| 亲戚 | qīnqi | relatives |

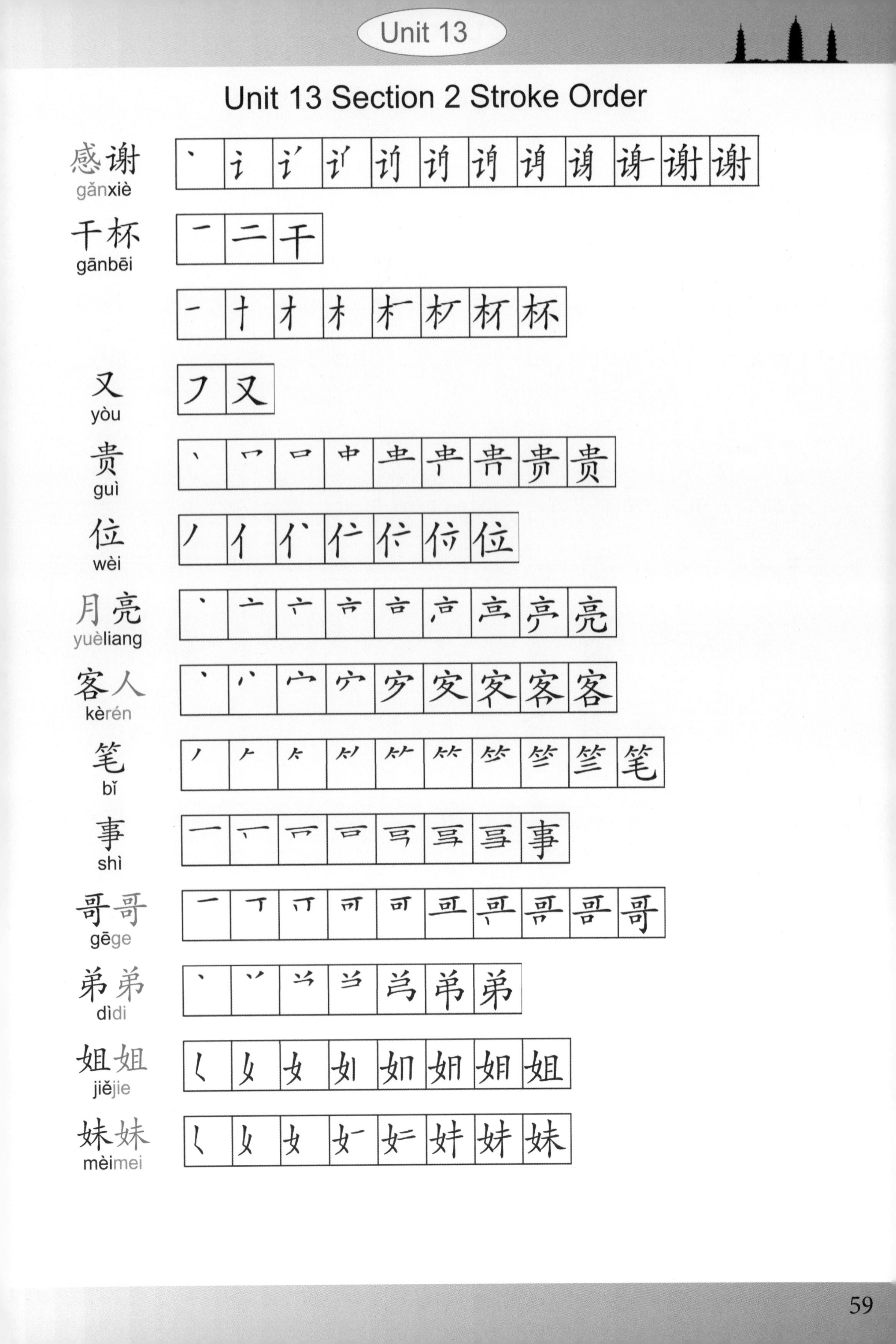

# Unit 13 Section 2 Stroke Order

感谢 gănxiè

干杯 gānbēi

又 yòu

贵 guì

位 wèi

月亮 yuèliang

客人 kèrén

笔 bǐ

事 shì

哥哥 gēge

弟弟 dìdi

姐姐 jiějie

妹妹 mèimei

## Section 2 – Customs

Activity 13.16

Practise writing the following characters:

谢

干

杯

又

贵

位

亮

客

笔

事

哥

弟

姐

妹

### Activity 13.17
Draw lines to match the situation with the appropriate exclamation.

| | |
|---|---|
| 你朋友给你生日礼物 | "别走了！" |
| 你和家人一起过年 | "干杯！" |
| 你在饭馆庆祝朋友的生日 | "我迷路了！" |
| 你朋友快要回家去了 | "谢谢！" |
| 你在路上不知道怎么走 | "新年快乐！" |

### Activity 13.18
Split up the following words and write each under the correct heading according to the tone of each character. As an example, 感 has been completed for you.

感谢　以为　干杯　月亮　文化

| 1st tone | 2nd tone |
|---|---|
| | |

| 3rd tone | 4th tone |
|---|---|
| 感 | |

### Activity 13.19
Draw a line to match each Chinese word with its English translation.

| | |
|---|---|
| 礼貌 | guest |
| 邀请 | polite |
| 亲戚 | relatives |
| 客人 | to invite |
| 感谢 | to understand |
| 明白 | thanks |

### Activity 13.20
Write the pinyin and tones of these sentences.

1. 虽然他在上海只有两个月，可是他交了很多朋友。

2. 哥哥以为北京不会下雪。

3. 过年的时候，我的亲戚会来看我。

4. 坐火车又便宜又舒服。

5. 他邀请你去看他的比赛。你想不想去？我们一起去吧。

### Activity 13.21
Below, fill in the blanks with appropriate words.

1. ☐人来了没有？

2. 来，大家☐☐！

3. ☐☐离开的时☐，对父亲说"祝你身体☐☐！"

4. ☐☐明天没有空。他有☐。

| | | | |
|---|---|---|---|
| 候 | 健康 | 弟弟 | 女儿 |
| 客 | 干杯 | 事 | |

## Activity 13.22

Translate the following sentences into Chinese.

1. Chinese culture and British culture are very different.

2. You've given me such a beautiful gift! Thank you very much!

3. May I ask, what's your name?

4. I don't understand how traffic can be this bad in the morning!

5. During the Mid-Autumn Festival, my family looks at the moon and eats mooncakes.

## Activity 13.23

Match the start and end of each sentence to make a logical complete sentence.

| | |
|---|---|
| 这个人真好， | 又去了长城，又学了汉语。 |
| 这儿的天气很好， | 又暖和又不太热。 |
| 我游泳的时候， | 她又聪明又努力。 |
| 她今天很不舒服， | 又感冒，又伤了她的腿。 |
| 他的假期过得很快乐， | 觉得又高兴又累。 |

## Activity 13.24

Below, write what you like to do in each of the seasons. Write in complete sentences.

1. 春天的时候，我喜欢……

2. 夏天

3. 秋天

4. 冬天

5. 你怎么过生日？

### Activity 13.25
Read the passage about cultural differences and answer the questions in English.

虽然英国人和美国人都说英语，可是他们的文化和生活习惯很不一样。英国人经常会在咖啡馆里喝茶。可是很多美国人买咖啡拿走，一边走路，一边喝咖啡。美国人经常对陌生人说"早上好！"可是，很多英国人不会这样。英国人经常坐公共汽车，可是美国的公共汽车很少。很多人说英国人比美国人礼貌。

陌生人 mòshēngrén = stranger

1. According to this passage, what do many British people do in a cafe?
2. How do Americans drink coffee?
3. How do Americans approach strangers?
4. What comment is made about public transport?
5. What comparison is made in the final sentence?

### Activity 13.26
Translate the following passage into English.

中国新年是我最喜欢的节日。因为我们的亲戚都住在很远的地方，所以他们不能来过年。我们每年请朋友来和我们一起庆祝。我们每年去公园看灯笼和舞狮。路上有很多人卖鞭炮，我们有时候会放鞭炮。

### Activity 13.27
Translate the following passage into Chinese.

When it was my older sister's birthday, I went to the store to buy her a gift. I didn't know what I should buy. She already has everything. She says that presents aren't important, but I still wanted to get her something. In the end, I bought her a pen.

# Unit 14 Section 1 Vocabulary

| | | |
|---|---|---|
| 微信 | Wēixìn | WeChat |
| 字典 | zìdiǎn | dictionary |
| 新闻 | xīnwén | news |
| 网站 | wǎngzhàn | website |
| 笔友 | bǐyǒu | pen pal |
| 网友 | wǎngyǒu | online friend, internet user |
| 电子邮件 | diànzǐ yóujiàn | email |
| 找 | zhǎo | to look for |
| 从来不 | cóngláibù | never |
| 如果, 要是 | rúguǒ, yàoshi | if |
| 网页 | wǎngyè | webpage |
| 复制 | fùzhì | to copy |
| 下载 | xiàzǎi | to download |
| 打字 | dǎ zì | to type |
| 删除 | shānchú | to delete |
| 怎么办 | Zěnme bàn? | What can be done? |
| 担心 | dānxīn | to worry |

**Online vocabulary**

| | | |
|---|---|---|
| 博客 | bókè | blog |
| 清楚 | qīngchu | clear |
| 风险 | fēngxiǎn | risk |
| 事儿 | shìr | matter, thing |
| 怕 | pà | afraid |
| 开始 | kāishǐ | to start |
| 到底 | dàodǐ | after all |

# Unit 14 Section 1 Stroke Order

新闻 xīnwén

邮件 yóujiàn

找 zhǎo

如果 rúguǒ

下载 xiàzǎi

清楚 qīngchu

怕 pà

开始 kāishǐ

忙 máng

电脑 diànnǎo

越 yuè

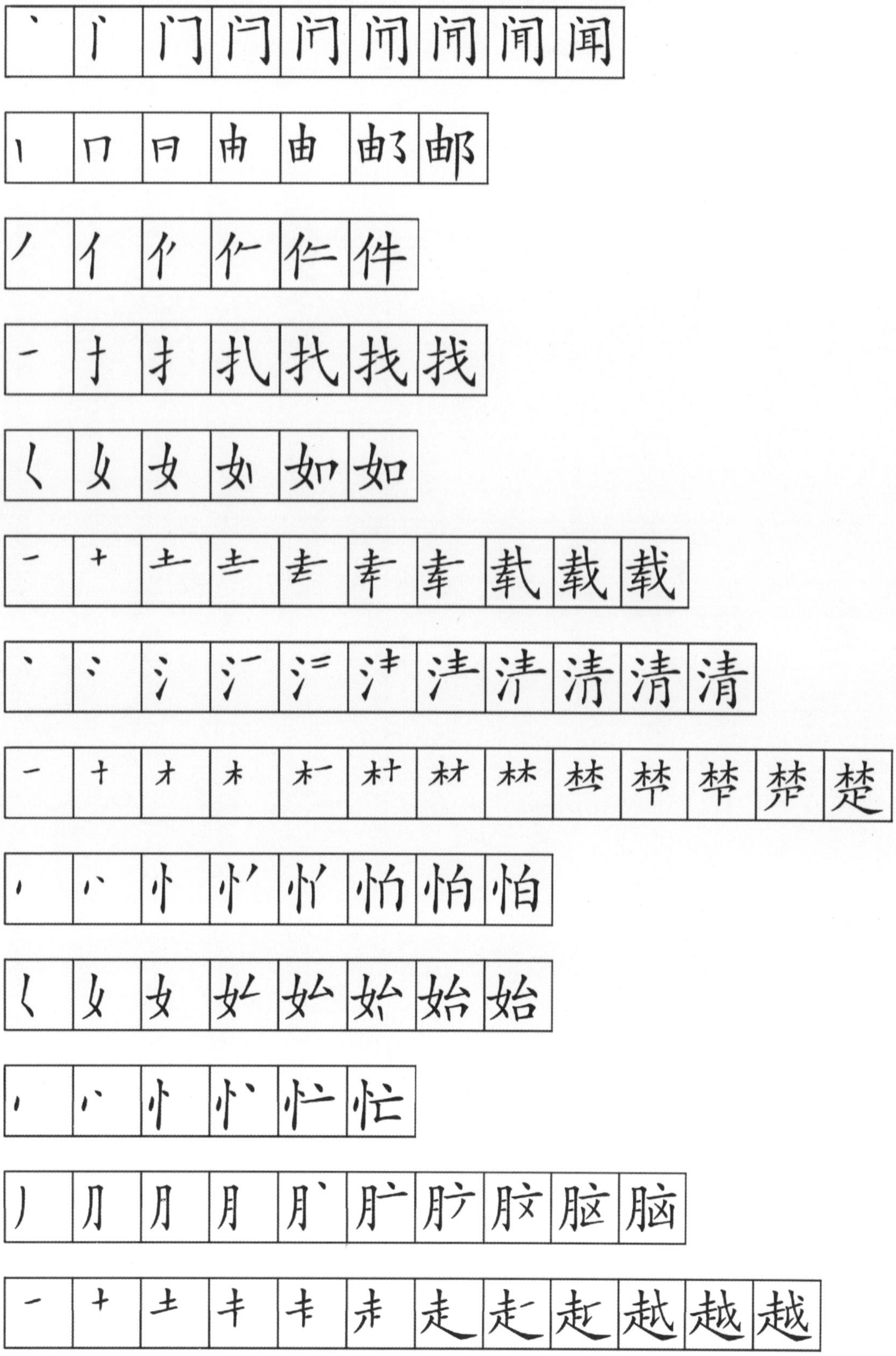

# 上网 Online Life
## Section 1 – Internet

Activity 14.1
Practise writing the following characters:

| | | | | | | | | | | | |
|---|---|---|---|---|---|---|---|---|---|---|---|
| 闻 | | | | | | | | | | | |
| 邮 | | | | | | | | | | | |
| 件 | | | | | | | | | | | |
| 找 | | | | | | | | | | | |
| 如 | | | | | | | | | | | |
| 载 | | | | | | | | | | | |
| 清 | | | | | | | | | | | |
| 楚 | | | | | | | | | | | |
| 怕 | | | | | | | | | | | |
| 始 | | | | | | | | | | | |
| 忙 | | | | | | | | | | | |
| 脑 | | | | | | | | | | | |
| 越 | | | | | | | | | | | |

### Activity 14.2

Draw lines to match each situation with the appropriate resource.

| | |
|---|---|
| 你想知道最近的事儿 | 字典 |
| 你想知道明天的天气 | 微信 |
| 你想和朋友聊天 | 天气预报 |
| 你想知道一个汉字的意思 | 教育网站 |
| 你想在网上学习新的科目 | 新闻 |

### Activity 14.3

Split up the following words and write each under the correct heading according to the tone of each character. As an example, 新 has been completed for you.

新闻　　从来　　开始　　清楚　　打字

| 1st tone | 2nd tone |
|---|---|
| 新 | |
| **3rd tone** | **4th tone** |
| | |

### Activity 14.4

Draw a line to match each Chinese word with its English translation.

| | |
|---|---|
| 到底 | website |
| 复制 | after all |
| 网页 | risk |
| 网站 | to delete |
| 删除 | to copy |
| 担心 | webpage |
| 风险 | to worry |

### Activity 14.5

Write the pinyin and tones of these sentences.

1. 弟弟每天都用微信和朋友聊天儿。

2. 他又是丈夫，又是父亲。

3. 我不认识这个汉字，可是我没有字典。

4. 你到底想去哪儿过圣诞节？

5. 那个字我打错了。可以删除吗？

### Activity 14.6

Below, fill in the blanks with appropriate words.

| | | | | |
|---|---|---|---|---|
| 如果 | 笔友 | 冬天 | 参加 | |
| 怕 | 找 | 别 | 戴 | 药 |

1. 我没有☐☐。你呢？

2. 你的课本☐到了没有？

3. ☐担心，你吃☐就会好了。

4. ☐☐你太累了，就不用☐☐比赛了。

5. 他☐冷。还没到☐☐，他已经☐戴帽子。

## Activity 14.7

Translate the following sentences into Chinese.

1. He never writes me emails!

2. You can look for a map online.

3. Have you ever downloaded a movie? I don't know how it's done.

4. If you don't read the news, you won't know anything.

5. Can you see the computer? You can see it clearly from here.

## Activity 14.8

Match the start and end of each sentence to make a logical complete sentence.

| | |
|---|---|
| 糟糕！我删除了 | 很有意思。 |
| 那位作家又写书 | 担心呢？ |
| 你别下载 | 老师的电子邮件！ |
| 你怎么这么 | 又写博客。 |
| 来，看一下这个网页， | 音乐。 |

## Activity 14.9

Below, finish each sentence to make an 'if-then' statement.

1. 如果你不努力学习……

2. 如果明天的天气很热……

3. 要是你不喜欢吃肉……

4. 如果我能去体育馆……

5. 要是你忘了带护照……

## Activity 14.10

Read the passage about Julie learning to use a computer answer the questions in English.

我九岁的时候开始学电脑。我们家不富有，没有钱买电脑。可是我小镇的图书馆设备很好，电脑又多又新。我每天放学以后会去图书馆用电脑。一天，有一位老太太来了。她说她要教我怎么打字。我怕自己给她麻烦，可是她说没事儿，因为她有空，所以来图书馆做义工。她说她很喜欢帮助年轻人学习。

1. How old was Julie when she started using a computer?
2. What does she say about the library's computers?
3. When did she usually go to the library?
4. Who taught Julie to type?
5. Why was she feeling anxious?

## Activity 14.11

Answer the questions for yourself in Chinese. Use complete sentences.

联系 liánxì = to connect

1. 你下载过电影吗？
2. 你怎么和朋友联系？你用过微信吗？
3. 你觉得小孩子从几岁开始学电脑好？
4. 你喜欢跟什么样的人交朋友？
5. 如果朋友觉得担心，你会对他说什么？

## Activity 14.12

Translate the following passage into English.

你的医生说你要多运动。可是如果你的工作太忙，没有空，你怎么办？别担心了，你不用出门，也能在家里做运动！每天跳舞对身体有好处。网上有不少的舞蹈短片。如果你不会跳舞，就可以上网找短片学！

舞蹈短片 wǔdǎo duǎnpiàn = dance video

## Activity 14.13

Translate the following passage into Chinese.

I always read the news, but my younger brother never does. He says that he is too busy. He only has a little free time, and believes (mistakenly) that emailing his friends is more important. I don't understand. I should speak clearly to him.

# Unit 14 Section 2 Vocabulary

| | | |
|---|---|---|
| 越来越 | yuè lái yuè | more and more |
| 网民 | wǎngmín | internet user |
| 好处 | hǎochu | advantage, benefit |
| 发 | fā | to send |
| 寄信 | jì xìn | to post a letter |
| 短信 | duǎnxìn | text message |
| 马上 | mǎshàng | immediately |
| 生意 | shēngyì | business |
| 坏处 | huàichu | disadvantage |
| 花 | huā | to spend (money, time) |
| 相信 | xiāngxìn | to believe |
| 大人 | dàrén | adult |
| 影响 | yǐngxiǎng | influence, to influence |
| 特别 | tèbié | special, especially |
| 省钱 | shěng qián | to save money |
| 不错 | búcuò | not bad |
| 公分 | gōngfēn | centimetre |
| 奇怪 | qíguài | strange |
| 左右 | zuǒyòu | more or less |

**Online vocabulary**

| | | |
|---|---|---|
| 谈恋爱 | tán liàn'ài | to date |
| 社会 | shèhuì | society |
| 约会 | yuēhuì | appointment, to arrange to meet |
| 不必 | búbì | not necessary |
| 罪行 | zuìxíng | crime |
| 有害 | yǒuhài | harmful |

# Unit 14 Section 2 Stroke Order

好处
hǎochu

坏处
huàichu

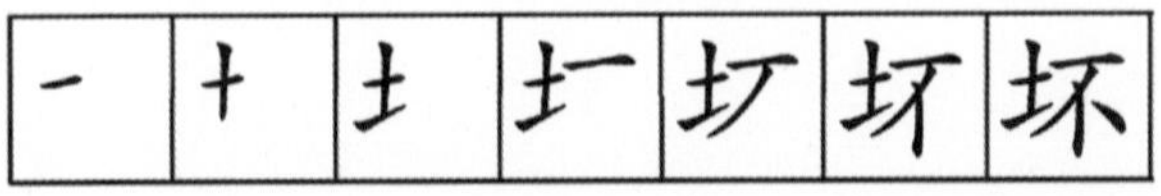

寄信
jìxìn

短信
duǎnxìn

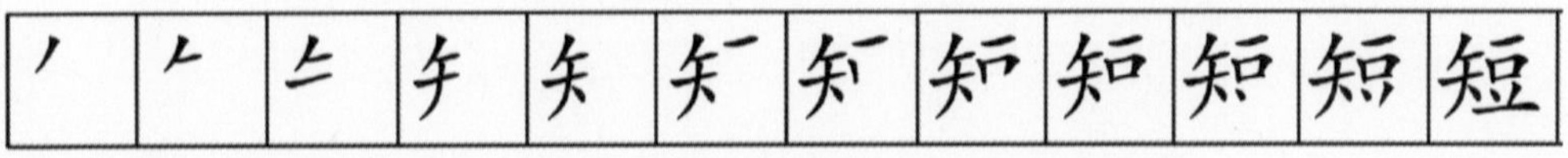

相信
xiāngxìn

影响
yǐngxiǎng

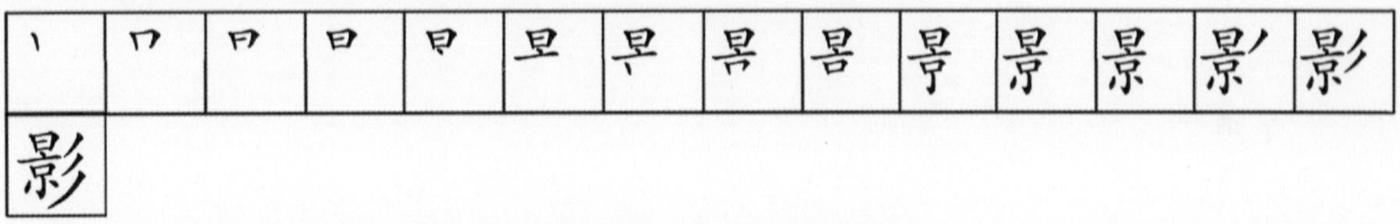

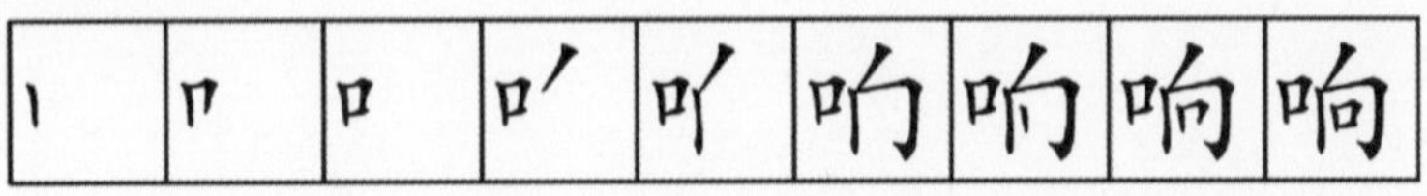

特别
tèbié

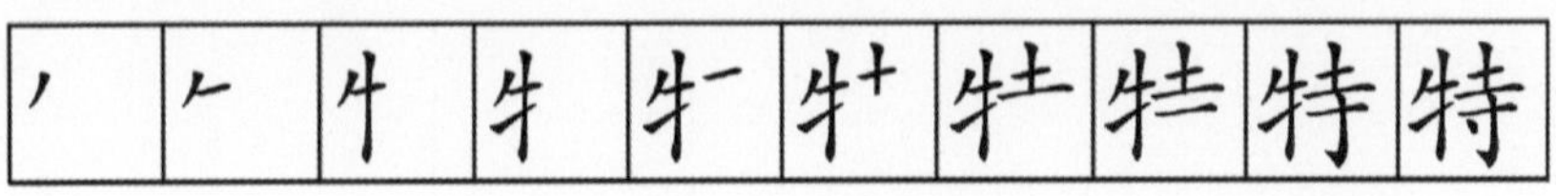

省钱
shěngqián

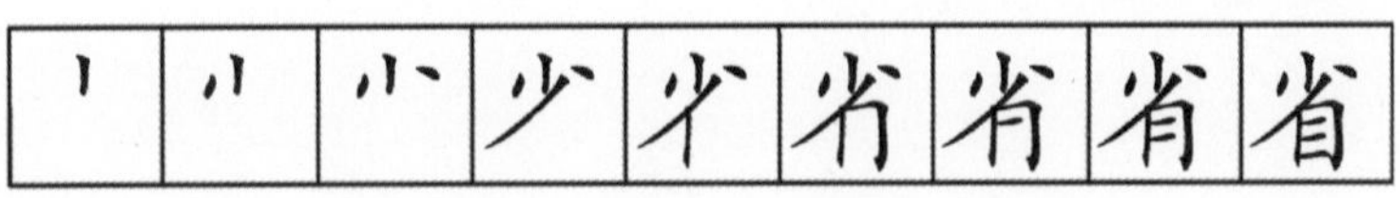

社会
shèhuì

不必
búbì

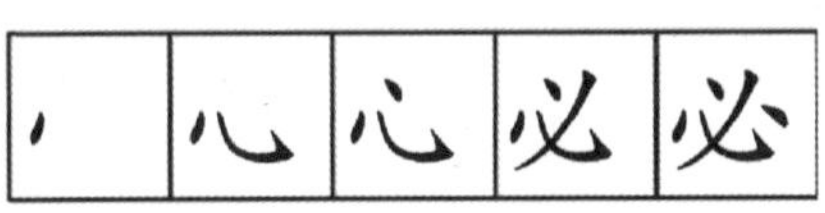

罪行
zuìxíng

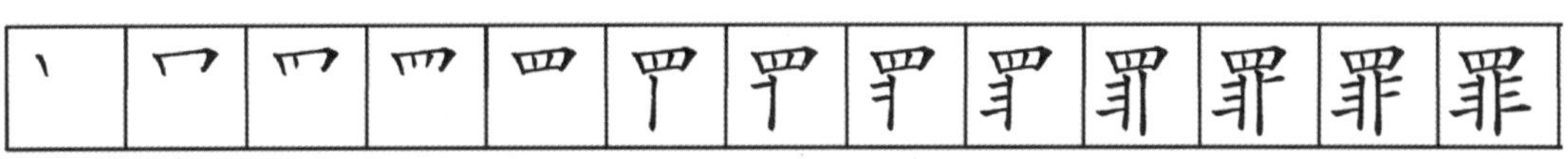

有害
yǒuhài

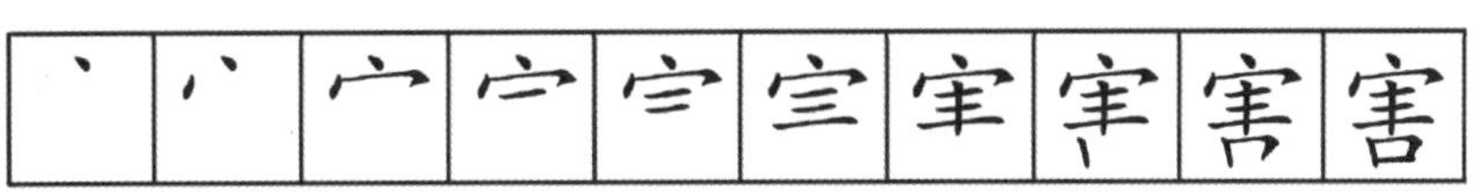

考试
kǎoshì

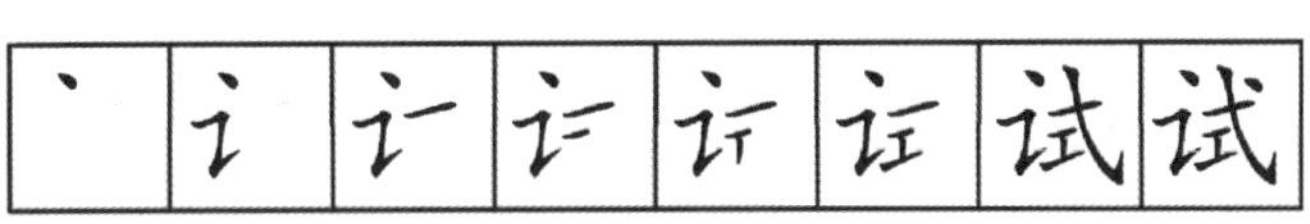

# Section 2 – Social trends

Activity 14.14
Practise writing the following characters:

处
坏
寄
短
相
影
响
特
省
社
必
罪
害
试

### Activity 14.15

Draw lines to match and make logical phrases.

| | |
|---|---|
| 花 | 上课 |
| 马上 | 很长时间 |
| 做 | 有意思 |
| 特别 | 左右 |
| 十三公分 | 生意 |

### Activity 14.16

Split up the following words and write each under the correct heading according to the tone of each character. As an example, 马 has been completed for you.

马上　　寄信　　省钱　　罪行　　影响

| 1st tone | 2nd tone |
|---|---|
| | |
| **3rd tone** | **4th tone** |
| 马 | |

### Activity 14.17

Draw a line to match each Chinese word with its English translation.

| | |
|---|---|
| 奇怪 | harmful |
| 网民 | to believe |
| 坏处 | strange |
| 约会 | appointment, to arrange to meet |
| 有害 | disadvantage |
| 相信 | text message |
| 短信 | web user |

### Activity 14.18

Write the pinyin and tones of these sentences.

1. 我不相信她的话。
2. 天气越来越凉快。
3. 他妹妹从来不上网做生意。
4. 我觉得这个网页特别奇怪。
5. 每天花很长时间看电视对身体有害。

### Activity 14.19

Below, fill in the blanks with appropriate words.

1. 你给我☐☐☐说在哪儿约会，好吗？
2. 那位老师对他的☐☐很大！
3. 你为什么☐来越不高兴？
4. ☐☐公司的好处比☐☐多。
5. 他的物理☐☐考得☐☐。

| | | |
|---|---|---|
| 越 | 影响 | 离开 |
| 考试 | 不错 | 发 |
| 坏处 | 短信 | |

## Activity 14.20

Translate the following sentences into Chinese.

1. Grandad never uses email. He always sends letters.

2. Downloading this music isn't a crime.

3. Please believe me!

4. Immediately after eating lunch, she'll take her chemistry exam.

5. Spending more than four hours using the computer is bad for your health.

## Activity 14.21

Match the start and end of each sentence to make a logical complete sentence.

| | |
|---|---|
| 姐姐给你的礼物 | 对学习不好。 |
| 你越说 | 社会最大的问题。 |
| 有人说谈恋爱 | 很特别。 |
| 真奇怪，他从来没跟我 | 说到这件事儿。 |
| 奶奶说毒品是 | 我越不相信你 。 |

## Activity 14.22

Below are four situations. Write a 好处 and 坏处 for each.

1. 住在大城市里

2. 用手写字（不打字）

3. 有特别多的爱好

4. 总是爱省钱

Some useful words: 罪行，安静，交通，慢，清楚，快，好看，忙，有意思，快乐，好玩儿，生活，贵，便宜，从来不，从来没

## Activity 14.23

Read the passage about Jody's business and answer the questions in English.

我特别喜欢做工艺。我从小就开始做自己的衣服。我还会给弟弟做衣服。两年前，一个朋友说我应该上网做生意，开网上的衣服店。那时候我不知道怎么做，也不相信自己。可是我看了一些生意书，上网找了教学视频。去年我开了网上的衣服店。我可以在家卖自己做的衣服。这份工作又舒服又能赚钱！

教学视频 jiàoxué shìpín = instructional video

1. For how long has Jody been making clothes?

2. What was her friend's suggestion?

3. How did Jody initially respond to this advice?

4. What does she like about working at home? Give two details.

## Activity 14.24

Answer the questions for yourself in Chinese. Use complete sentences.

1. 你每天花多长时间上网？

2. 发电子邮件、打电话、发短信：你最喜欢哪个？为什么？

3. 什么事儿让你很快乐？

4. 你怎么省钱？

5. 什么科目对你最重要？为什么？

## Activity 14.25

Translate the following passage into English.

网络很有用，可是也有坏处。去年，我哥哥上瘾了网络游戏。他很喜欢跟朋友玩儿游戏，每天玩儿很长时间。爸爸妈妈很担心，因为游戏影响了他的学习。最后，爸爸妈妈不让他玩儿游戏。哥哥很伤心，可是他也知道这个游戏对他不好。一个月以后，哥哥删除了这个游戏，再也不想玩儿了。

## Activity 14.26

Translate the following passage into Chinese.

The day before yesterday my younger brother took a history exam. I sent him a text message to ask him how it went. He called me immediately. He said, "It was really difficult. I didn't know what I should write, so I didn't write anything." I hope the teacher won't be angry.

# Unit 15 Section 1 Vocabulary

| | | |
|---|---|---|
| 警察 | jǐngchá | police officer |
| 别人 | biérén | other people |
| 压力 | yālì | pressure |
| 职业 | zhíyè | profession |
| 厨师 | chúshī | chef |
| 科学家 | kēxuéjiā | scientist |
| 毕业 | bìyè | to graduate |
| 赚钱 | zhuàn qián | to earn money |
| 商场 | shāngchǎng | marketplace |
| 售货员 | shòuhuòyuán | salesperson |
| 工资 | gōngzī | salary, pay |
| 新加坡 | Xīnjiāpō | Singapore |
| 名片 | míngpiàn | business card |
| 听说 | tīngshuō | one hears (that) |
| 差 | chà | inferior, lacking |
| 流利 | liúlì | fluent |
| 同事 | tóngshì | colleague |
| 试 | shì | to try |
| 申请 | shēnqǐng | to apply |
| 性别 | xìngbié | gender, sex |
| 出生 | chūshēng | to be born |
| 日期 | rìqī | date |
| 联系 | liánxì | to contact |
| 地址 | dìzhǐ | address |

**Online vocabulary**

| | | |
|---|---|---|
| 家庭主妇 | jiātíngzhǔfù | housewife |
| 打算 | dǎsuàn | to plan to, to intend |
| 背景 | bèijǐng | background |
| 年纪 | niánjì | age |
| 经验 | jīngyàn | experience |
| 失业 | shīyè | unemployment, to lose one's job |
| 希望 | xīwàng | hope, to hope |
| 为了 | wèile | for the sake of |
| 然后 | ránhòu | afterwards |

# Unit 15 Section 1 Stroke Order

压力
yālì

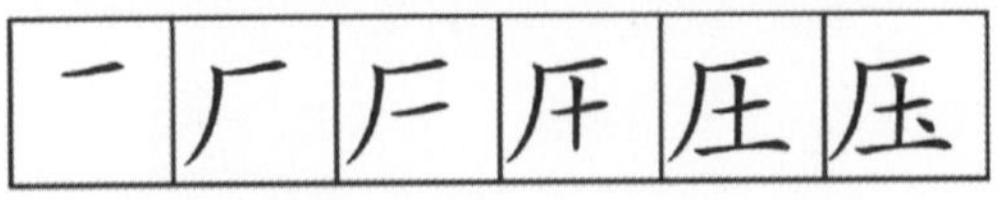

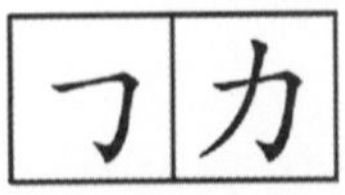

厨师
chúshī

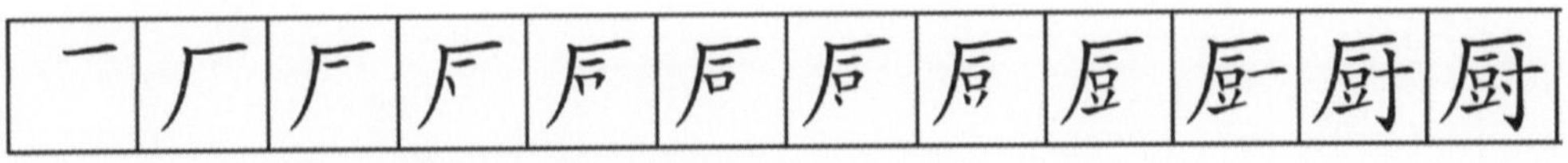

毕业
bìyè

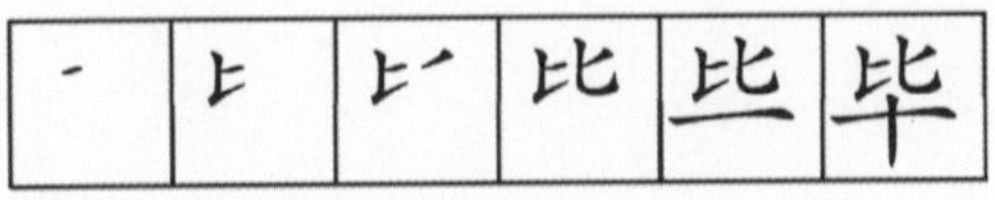

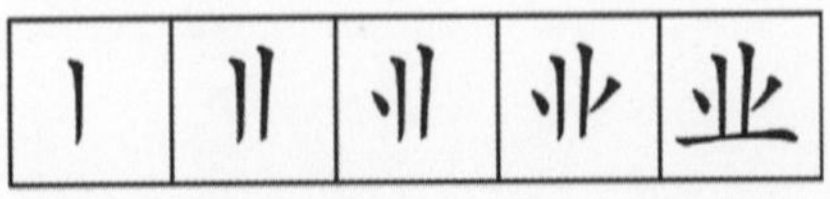

赚钱
zhuànqián

工资
gōngzī

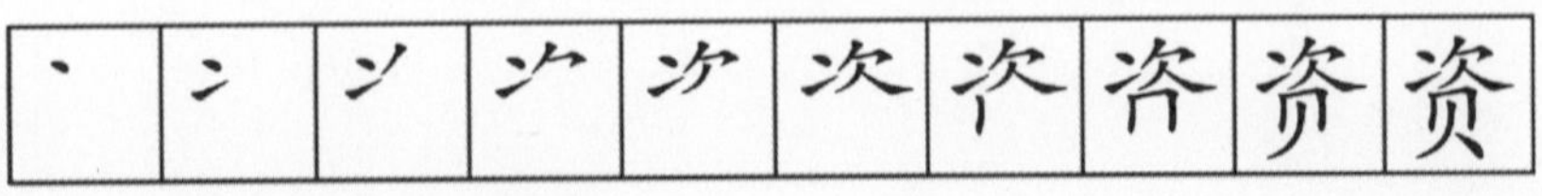

名片
míngpiàn

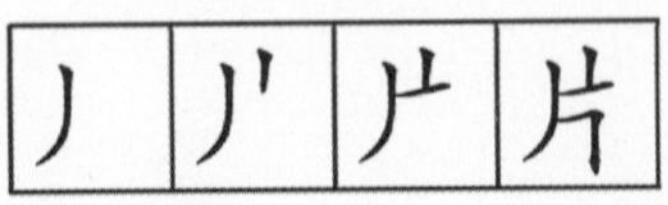

流利
liúlì

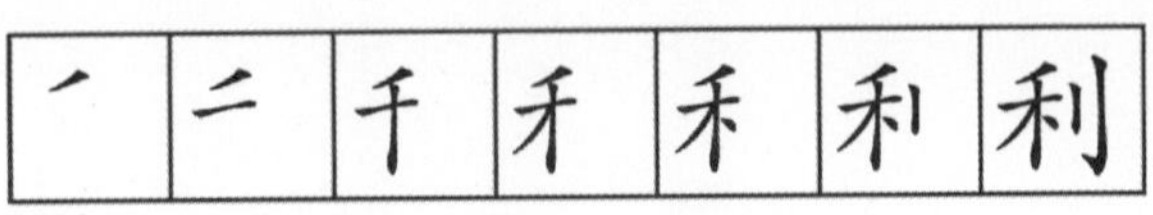

申请
shēnqǐng

联系
liánxì

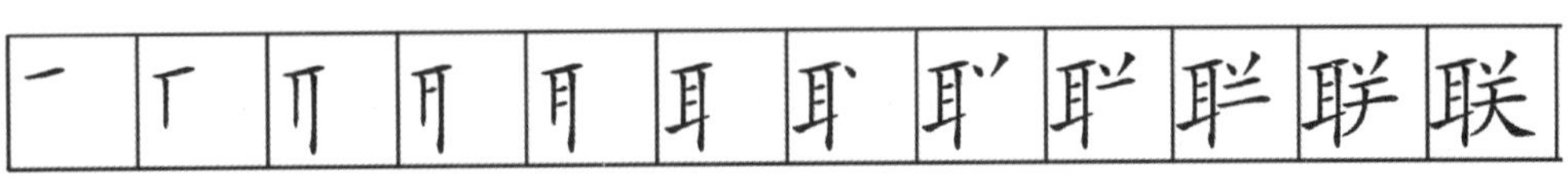

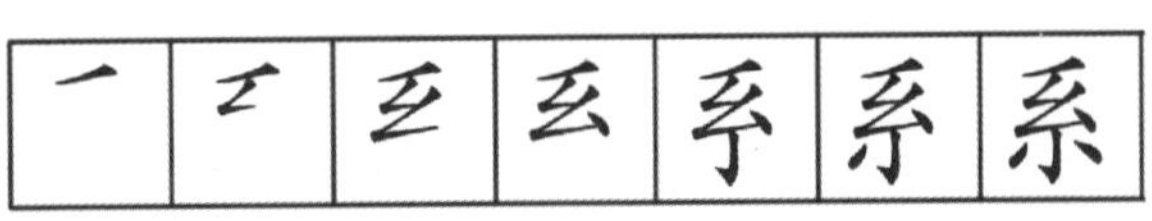

差
chà

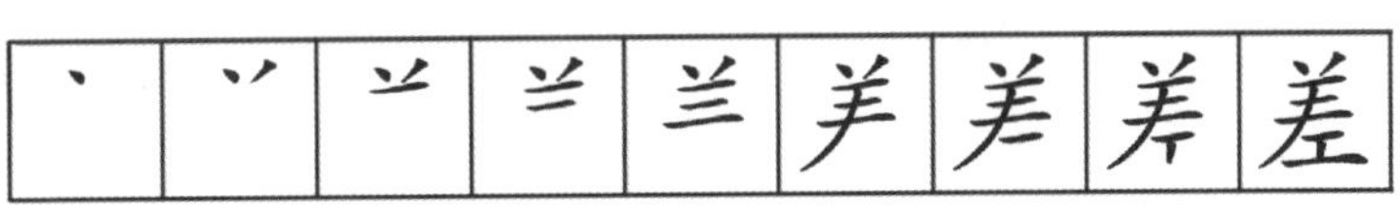

# 上班 Going to work

## Section 1 – Career plans

Activity 15.1

Practise writing the following characters:

压

力

厨

毕

业

赚

资

片

流

利

申

联

系

差

### Activity 15.2
Draw lines to match pairs and make logical phrases.

| | |
|---|---|
| 申请 | 工资 |
| 大学 | 办法 |
| 很高的 | 工作 |
| 联系 | 日期 |
| 出生 | 毕业 |

### Activity 15.3
Split up the following words and write each under the correct heading according to the tone of each character. As an example, 名 has been completed for you.

名片　流利　商场　压力　赚钱

| 1st tone | 2nd tone |
|---|---|
| | 名 |
| **3rd tone** | **4th tone** |
| | |

### Activity 15.4
Draw a line to match each Chinese word with its English translation.

| | |
|---|---|
| 职业 | to hope |
| 经验 | to plan |
| 性别 | profession |
| 打算 | police |
| 希望 | experience |
| 背景 | background |
| 警察 | gender, sex |

### Activity 15.5
Write the pinyin and tones of these sentences.

1. 我听说新加坡人的工作很忙。
2. 他是厨师，你什么菜都会做。
3. 他的英文说得真流利！
4. 你的工作背景是什么？

### Activity 15.6
Below, fill in the blanks with appropriate words.

工资　同事
赚钱　压力　丢
名片　科学家

1. 我的同学很富有，他不用找工作＿＿＿＿。
2. 父母给我的＿＿＿＿太大了。
3. 糟糕！我＿＿了他的＿＿＿＿！
4. ＿＿＿＿＿＿的＿＿＿＿比售货员的高得多。
5. 你不要听＿＿＿＿的话。他们没有你的经验。

## Activity 15.7

Translate the following sentences into Chinese.

1. She doesn't work in the mall. She only goes there to buy clothes.

2. He speaks Chinese fluently. He should do business in China.

3. My colleague thought (mistakenly) that the salary was higher.

4. When will you graduate?

5. Get in touch with me right away when you want to apply, okay?

## Activity 15.8

Match the start and end of each sentence to make a logical complete sentence.

| | |
|---|---|
| 我看得清楚 | 这杯可乐 |
| 我写不完 | 英文书 |
| 我走不到 | 艺术作业 |
| 我看得懂 | 四川饭馆 |
| 我喝不下 | 这张照片 |

## Activity 15.9

Fill in your personal details below in Chinese, or make up the details for a fictional person.

1. 姓名:
2. 出生日期:
3. 出生地点:
4. 性别:
5. 地址:
6. 电子邮件:
7. 国籍:

## Activity 15.10

Read the passage about Cecelia's mother and answer the questions in English.

我妈妈是厨师。她喜欢她的工作，可是因为厨师的工资不高，所以她下班以后还去做第二份工作。她晚上在商场当售货员卖茶。她每天工作很累，一天站了十四个小时。她回家后，马上就睡觉。她为什么这么努力工作？我知道，是为了我和弟弟。她经常说她希望我们的生活能更舒服。

1. What is Cecelia's mother's primary job?
2. What is her second job?
3. What does she do when she gets home?
4. Why does she work so hard?
5. What does she often tell Cecelia?

## Activity 15.11

Answer the questions for yourself in Chinese. Use complete sentences.

1. 你觉得工作中最重要的是什么？
2. 你最想试试什么样的工作？
3. 你觉得最容易申请的工作是什么？最难的呢？
4. 你觉得你现在的学习或工作压力大不大？为什么？

## Activity 15.12

Translate the following passage into English.

我在博物馆工作。因为我的同事快要走了， 所以最近我们找别人来这儿工作。昨天有人来申请，可是他很不合适。我请他给我名片的时候，他说“对不起，我没带”。他又没有经验，又没有专业背景。听说这个人两年以前就失业了。我希望他很快能找到合适的工作。

专业 zhuānyè = specialty

## Activity 15.13

Translate the following passage into Chinese.

Two weeks ago my older brother applied to work in a company. He wants to be a manager. When he was a student he spoke French fluently. However, after graduating university he didn't speak any French, so now his French is very poor. The company still hasn't gotten in touch with him. I said to him, "Don't worry, you can find a good job."

# Unit 15 Section 2 Vocabulary

| | | |
|---|---|---|
| 面试 | miànshì | interview |
| 面谈 | miàntán | interview |
| 个人 | gèrén | individual, personal |
| 简历 | jiǎnlì | CV, resume |
| 笑 | xiào | smile, to smile or laugh |
| 目标 | mùbiāo | goal, objective |
| 理想 | lǐxiǎng | ideal |
| 义工 | yìgōng | volunteer work, volunteer worker |
| 兼职 | jiānzhí | to do part-time (work) |
| 能力 | nénglì | ability |
| 机会 | jīhuì | opportunity |
| 广告 | guǎnggào | advertisement |

(No online vocabulary)

# Unit 15 Section 2 Stroke Order

打算 dǎsuàn

| | | | | | | | | | | | | | |
|---|---|---|---|---|---|---|---|---|---|---|---|---|---|
| ノ | 𠂉 | 𠂉 | ⺮ | ⺮ | ⺮ | ⺮ | 笱 | 笱 | 笛 | 笪 | 笪 | 箅 | 算 |

背景 bèijǐng

| | | | | | | | | |
|---|---|---|---|---|---|---|---|---|
| 丨 | 十 | 彐 | 彐 | 北 | 北 | 背 | 背 | 背 |

经验 jīngyàn

| | | | | | | | | | |
|---|---|---|---|---|---|---|---|---|---|
| ㇆ | 马 | 马 | 马 | 马 | 驴 | 验 | 验 | 验 | 验 |

希望 xīwàng

| | | | | | | |
|---|---|---|---|---|---|---|
| ノ | 乂 | 乂 | 㐅 | 𠚤 | 希 | 希 |

| | | | | | | | | | | |
|---|---|---|---|---|---|---|---|---|---|---|
| 丶 | 亠 | 亡 | 亡 | 㠩 | 朚 | 朚 | 望 | 望 | 望 | 望 |

面谈 miàntán

| | | | | | | | | | |
|---|---|---|---|---|---|---|---|---|---|
| 丶 | 讠 | 讠 | 讠 | 讠 | 讠 | 讠 | 谈 | 谈 | 谈 |

简历 jiǎnlì

| | | | | | | | | | | | | |
|---|---|---|---|---|---|---|---|---|---|---|---|---|
| ノ | 𠂉 | 𠂉 | ⺮ | ⺮ | ⺮ | ⺮ | ⺮ | 简 | 简 | 简 | 简 | 简 |

| | | | |
|---|---|---|---|
| 一 | 厂 | 万 | 历 |

笑 xiào

| | | | | | | | | | |
|---|---|---|---|---|---|---|---|---|---|
| ノ | 𠂉 | 𠂉 | ⺮ | ⺮ | ⺮ | ⺮ | 笁 | 笑 | 笑 |

目标 mùbiāo

| | | | | | | | | |
|---|---|---|---|---|---|---|---|---|
| 一 | 十 | 才 | 木 | 木 | 杧 | 标 | 标 | 标 |

义工 yìgōng

| | | |
|---|---|---|
| 丶 | 丿 | 义 |

兼职 jiānzhí

| | | | | | | | | | |
|---|---|---|---|---|---|---|---|---|---|
| 丶 | 丷 | 丷 | 兰 | 兰 | 兰 | 兼 | 兼 | 兼 | 兼 |

| | | | | | | | | | | |
|---|---|---|---|---|---|---|---|---|---|---|
| 一 | 𠀀 | 𠃍 | 月 | 月 | 耳 | 耳 | 职 | 职 | 职 | 职 |

广告 guǎnggào

| | | | | | | |
|---|---|---|---|---|---|---|
| 丿 | 𠂉 | 牛 | 生 | 告 | 告 | 告 |

## Section 2 – Applying for a job

Activity 15.16
Practise writing the following characters:

| | | | | | | | | | | | |
|---|---|---|---|---|---|---|---|---|---|---|---|
| 算 | | | | | | | | | | | |
| 背 | | | | | | | | | | | |
| 验 | | | | | | | | | | | |
| 希 | | | | | | | | | | | |
| 望 | | | | | | | | | | | |
| 谈 | | | | | | | | | | | |
| 简 | | | | | | | | | | | |
| 历 | | | | | | | | | | | |
| 笑 | | | | | | | | | | | |
| 标 | | | | | | | | | | | |
| 义 | | | | | | | | | | | |
| 兼 | | | | | | | | | | | |
| 职 | | | | | | | | | | | |
| 告 | | | | | | | | | | | |

### Activity 15.17
Draw lines to match the words with their rough definitions.

| | |
|---|---|
| 理想 | 不上学了 |
| 义工 | 一张纸，上面写着你的工作经验 |
| 毕业 | 免费工作 |
| 面试 | 不能更好的事儿 |
| 简历 | 申请工作的时候，一对一的当面考试 |

### Activity 15.18
Split up the following words and write each under the correct heading according to the tone of each character. As an example, 面 has been completed for you.

面谈　　简历　　兼职　　广告　　能力

| 1st tone | 2nd tone |
|---|---|
| | |

| 3rd tone | 4th tone |
|---|---|
| | 面 |

### Activity 15.19
Draw a line to match each Chinese word with its English translation.

| | |
|---|---|
| 机会 | experience |
| 个人 | opportunity |
| 经验 | goal, objective |
| 地址 | individual, personal |
| 义工 | CV, resume |
| 简历 | address |
| 目标 | volunteer work, volunteer |

### Activity 15.20
Write the pinyin and tones of these sentences.

1. 你带了个人简历吗？
2. 当一个好父亲是他的目标。
3. 你理想的工作是什么？
4. 别笑了，我真的想做警察！
5. 经理一看我的名片，马上就笑了。

### Activity 15.21
Below, fill in the blanks with appropriate words.

1. 我＿＿＿＿那家公司会请我来参加＿＿＿＿。
2. 你以后＿＿＿＿申请什么样的工作？
3. 她没有＿＿＿＿，可是她很有＿＿＿＿。
4. 她有空的时候＿＿＿＿在公园做＿＿＿＿。
5. 你没有＿＿＿＿背景，为什么要＿＿老师？

| | |
|---|---|
| 经常 | |
| 希望 | 能力 |
| 教育 | 义工 |
| 做 | 经验 |
| 面试 | 打算 |

## Activity 15.22

Translate the following sentences into Chinese.

1. I think speaking to other people is easy. I'm not afraid of the interview.

2. Please take a look at this advertisement. What do you think of it?

3. Have you ever done volunteer work?

4. My colleague works part-time at the supermarket.

5. I never had the opportunity to go to China.

## Activity 15.23

Match the start and end of each sentence to make a logical complete sentence.

| | |
|---|---|
| 她参加了面谈，可是 | 个人简历很重要。 |
| 我很高兴， | 公司没有跟她联系。 |
| 做义工是 | 是早一年毕业，早一年开始工作。 |
| 我的理想 | 得到工作经验的好办法。 |
| 我觉得 | 明年我有机会学习法语。 |

## Activity 15.24

Your friend had a job interview several days ago. Using 是……的, write five questions you might ask her to find out details of the job (eg. who, where, what, when, and why).

1.

2.

3.

4.

5.

## Activity 15.25

Read the passage about Wang Fei's interview and answer the questions in English.

我是上个月从大学毕业，可是我五月就开始找工作了。找工作的压力很大。因为我没有很多钱，所以我一定要找好工作。我的朋友说我可以找义工，但是我怕工资不够。我的专业是生物学，我对动物感兴趣。我找自然保护区和动物园的工作。

愿意 yuànyì = to be willing

1. When did Wang Fei begin looking for work?

2. What is her current financial situation?

3. What is she afraid will happen if she does volunteer work?

4. What kind of job is she looking for?

## Activity 15.26

Answer the questions for yourself in Chinese. Use complete sentences.

1. 你参加过面试吗？为什么

2. 你觉得兼职工作的好处是什么？

3. 简历上应该写什么？

4. 你的学习目标是什么？

5. 什么事儿让你笑？

## Activity 15.27

Translate the following passage into English.

前天我去参加面试了。我很担心，吃不了早餐。我打算骑自行车去，因为自行车比公共汽车快。可是，我迷路了。交通也很不好，所以我晚了五分钟。面试过得很好，我很喜欢老板。老板跟我一样喜欢骑自行车。他说骑自行车对环境很好。

## Activity 15.28

Translate the following passage into Chinese.

My ideal job is to work in a school. I really like kids. Yesterday my older sister showed me a job ad for a new head of school. The school is in the suburbs. I don't have head of school experience, but I have ability. I hope I can attend an interview.

# Unit 16 Section 1 Vocabulary

| | | |
|---|---|---|
| 客厅 | kètīng | guest room |
| 灯 | dēng | light, lamp, lantern |
| 卧室 | wòshì | bedroom |
| 厨房 | chúfáng | kitchen |
| 卫生间 | wèishēngjiān | lavatory |
| 房子 | fángzi | house |
| 饭厅 | fàntīng | dining room |
| 有用 | yǒuyòng | useful |
| 洗衣机 | xǐyījī | washing machine |
| 洗 | xǐ | to clean, wash |
| 旁边 | pángbiān | next to |
| 洗手间 | xǐshǒujiān | lavatory |
| 厕所 | cèsuǒ | toilet |
| 房间 | fángjiān | room |
| 床 | chuáng | bed |
| 书架 | shūjià | bookshelf |
| 放 | fàng | to put |
| 着 | zhe | particle indicating an action is in progress |
| 空调 | kōngtiáo | air conditioning |
| 干净 | gānjìng | clean |
| 沙发 | shāfā | sofa |

**Online vocabulary**

| | | |
|---|---|---|
| 书房 | shūfáng | study |
| 花园 | huāyuán | garden |
| 浴室 | yùshì | bath room |
| 中间 | zhōngjiān | between |
| 冰箱 | bīngxiāng | refrigerator |
| 炉子 | lúzi | oven |

# Unit 16 Section 1 Stroke Order

客厅
kètīng

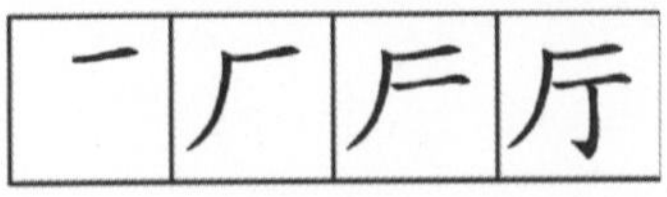

卧室
wòshì

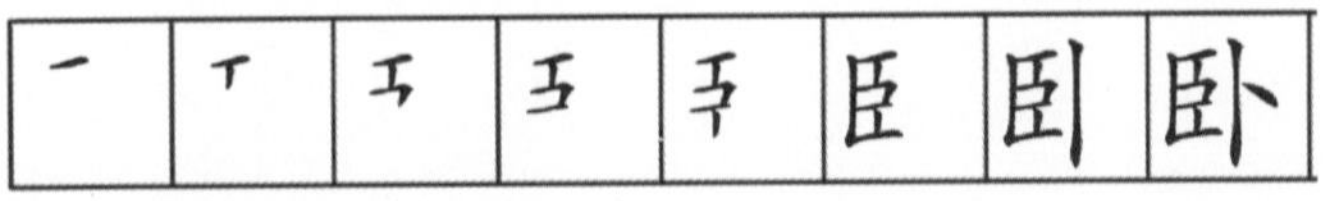

卫生间
wèishēngjiān

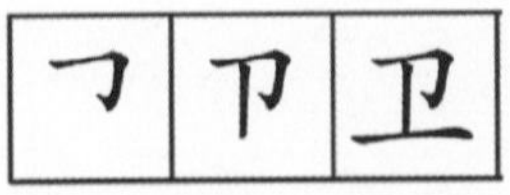

洗
xǐ

旁边
pángbiān

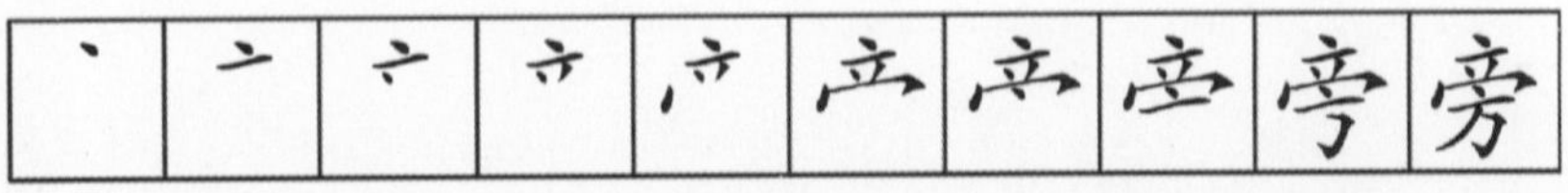

书架
shūjià

空调
kōngtiáo

干净
gānjìng

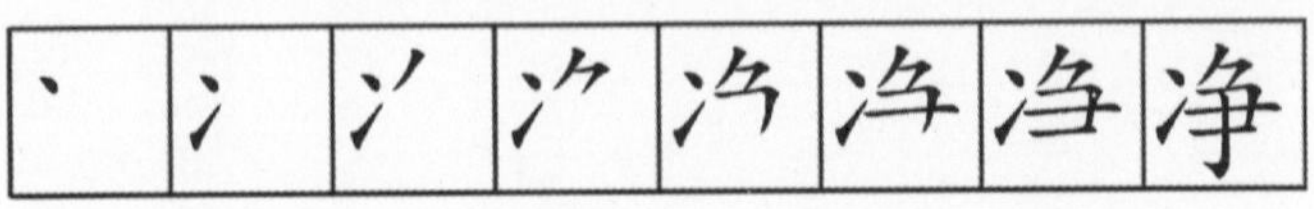

沙发
shāfā

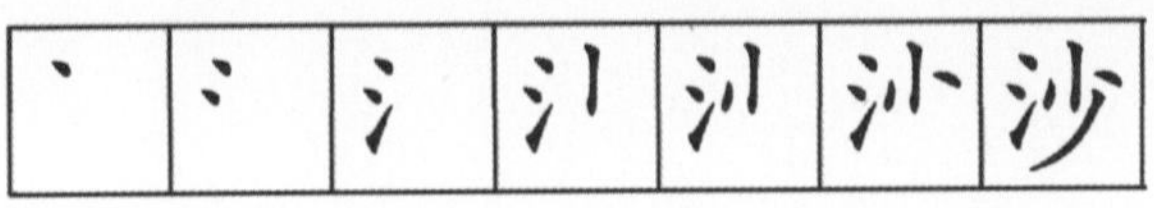

花园
huāyuán

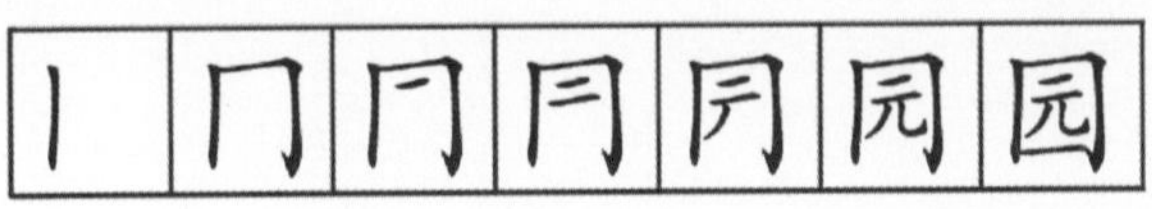

浴室
yùshì

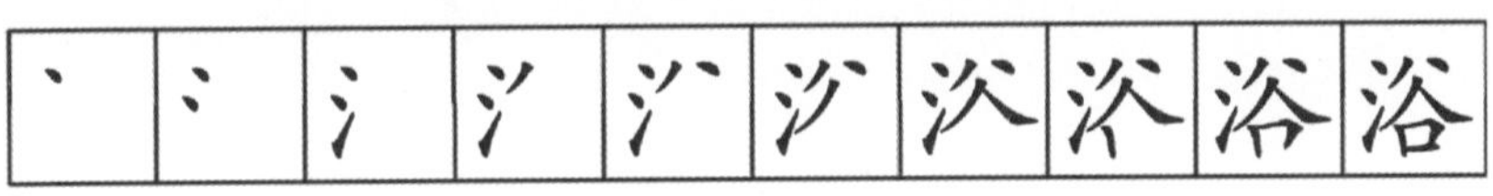

冰箱
bīngxiāng

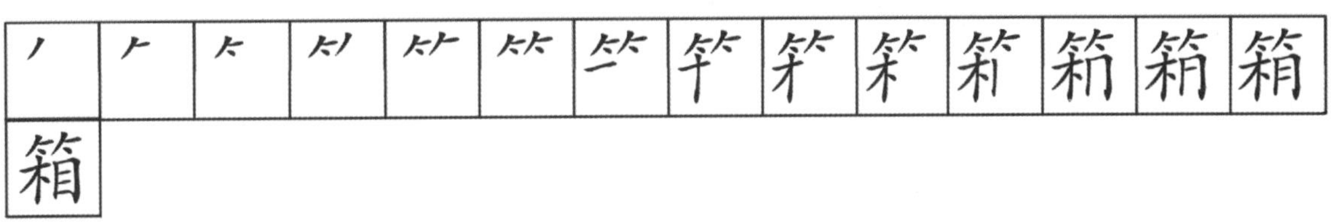

把
bǎ

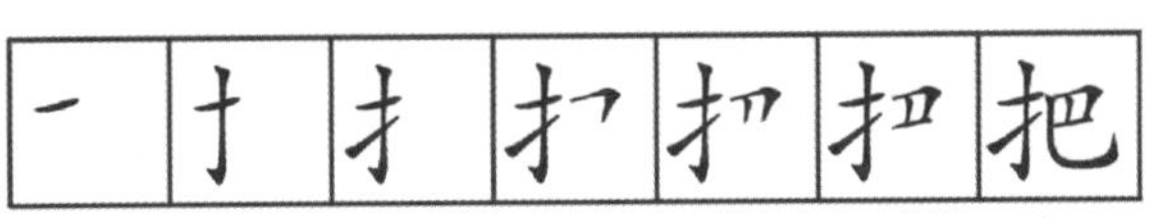

着
zhe

# 在家 Home Life

## Section 1 – My home

Activity 16.1

Practise writing the following characters:

| | | | | | | | | | | | |
|---|---|---|---|---|---|---|---|---|---|---|---|
| 厅 | | | | | | | | | | | |
| 卧 | | | | | | | | | | | |
| 卫 | | | | | | | | | | | |
| 洗 | | | | | | | | | | | |
| 旁 | | | | | | | | | | | |
| 架 | | | | | | | | | | | |
| 调 | | | | | | | | | | | |
| 净 | | | | | | | | | | | |
| 沙 | | | | | | | | | | | |
| 园 | | | | | | | | | | | |
| 浴 | | | | | | | | | | | |
| 箱 | | | | | | | | | | | |
| 把 | | | | | | | | | | | |
| 着 | | | | | | | | | | | |

### Activity 16.2
Draw lines to match pairs and make logical phrases.

| | |
|---|---|
| 开 | 电话 |
| 起 | 空调 |
| 洗 | 客厅 |
| 进 | 手 |
| 打 | 床 |

### Activity 16.3
Split up the following words and write each under the correct heading according to the tone of each character. As an example, 厨 has been completed for you.

厨房　书架　干净　冰箱　卧室

| 1st tone | 2nd tone |
|---|---|
| | 厨 |
| **3rd tone** | **4th tone** |
| | |

### Activity 16.4
Draw a line to match each Chinese word with its English translation.

| | |
|---|---|
| 沙发 | house |
| 卫生间 | oven |
| 浴室 | beside |
| 炉子 | room |
| 房子 | lavatory |
| 房间 | sofa |
| 旁边 | bath room |

### Activity 16.5
Write the pinyin and tones of these sentences.

1. 我把灯关上，你睡吧！

2. 妈妈把晚餐带出来了。

3. 我这儿有洗衣机，你不必用手洗衣服！

4. 请把书架放在沙发旁边，谢谢！

5. 哥哥是不是在花园里？

### Activity 16.6
Below, fill in the blanks with appropriate words.

1. ☐间很☐，请把☐☐关上，好吗？
2. 我昨天去了超市，☐☐里有很多苹果。
3. ☐☐☐怎么这么不☐☐呢？
4. 我看不☐☐，你帮我开☐好不好？
5. 我把书☐在床上，可是现在找不到。

| | |
|---|---|
| 卫生间 | 干净 |
| 冷 | 清楚 |
| 空调 | 冰箱 |
| 放 | 灯 |
| 房 | |

## Activity 16.7

Translate the following sentences into Chinese.

1. When did you buy your house?

2. The guest room is so beautiful!

3. I put the lamp next to the washing machine.

4. My little brother does homework in his bedroom every day.

5. The kitchen is cleaner than it was yesterday!

## Activity 16.8

Match the start and end of each sentence to make a logical complete sentence.

| | |
|---|---|
| 外边在下雪，我很冷， | 看着电视。 |
| 我的卧室里 | "厕所在哪儿？"是很有用的一句话。 |
| 我们的房子一共有 | 六个房间。 |
| 妹妹坐在沙发上 | 有很多书。 |
| 在外国旅游的时候， | 快开空调吧！ |

外国 wàiguó = foreign country

## Activity 16.9

Below are some objects. Using 把, write a sentence for each to state where in your home you would put each object.

1. 电脑

2. 电视机

3. 洗衣机

4. 冰箱

5. 空调

## Activity 16.10

Read the passage about Maria's home life and answer the questions in English.

我家很小。厨房是最小的房间。我的父亲做饭的时候，总是让我和弟弟去外边玩。他说如果不出去，房子就会太吵。只有姐姐不用出去玩。爸爸说她是高中生，可以在卧室里做作业。可是，我知道姐姐常常不会做作业。她现在坐在床上给朋友们发短信。

1. Which is the smallest room in Maria's house?
2. When does her father want the kids out of the house?
3. Which child is granted an exception, and why?
4. What does Maria think that sibling is doing right now?

## Activity 16.11

Answer the questions for yourself in Chinese. Use complete sentences.

1. 在你家，你最喜欢哪个房间？为什么？
2. 请说一下你理想中的房子怎么样。

## Activity 16.12

Translate the following passage into English.

我的房子很舒服：不大不小。我们住在小镇里，又安静又干净。这里的气候很舒服，四季如春。因为天气每天都很暖和，所以不用开空调。我的房子前边是一个大花园。我和姐姐常常在花园里打球、看书。

## Activity 16.13

Translate the following passage into Chinese.

My house has five rooms: a kitchen, a bathroom, a living room, and two bedrooms. My little sister and I live in one room. In the summer it is very hot because we don't have air conditioning. Sometimes I like to sleep in the garden.

# Unit 16 Section 2 Vocabulary

| | | |
|---|---|---|
| 遛狗 | liùgǒu | to walk a dog |
| 孙女 | sūnnǚ | granddaughter |
| 多大 | duōdà | how old |
| 孙子 | sūnzǐ | grandson |
| 比较 | bǐjiào | quite, rather, comparatively |
| 结婚 | jié hūn | to get married |
| 男朋友 | nánpéngyou | boyfriend |
| 婚礼 | hūnlǐ | wedding |
| 先 | xiān | first |
| 饥荒 | jīhuāng | famine |
| 坏 | huài | bad, broken |

**Online vocabulary**

| | | |
|---|---|---|
| 强 | qiáng | strong |
| 严肃 | yánsù | serious |
| 胡说 | húshuō | to talk nonsense |
| 告诉 | gàosu | to tell |
| 讨厌 | tǎoyàn | to dislike |
| 号码 | hàomǎ | number |
| 弱点 | ruòdiǎn | weak point |

# Unit 16 Section 2 Stroke Order

狗
gǒu

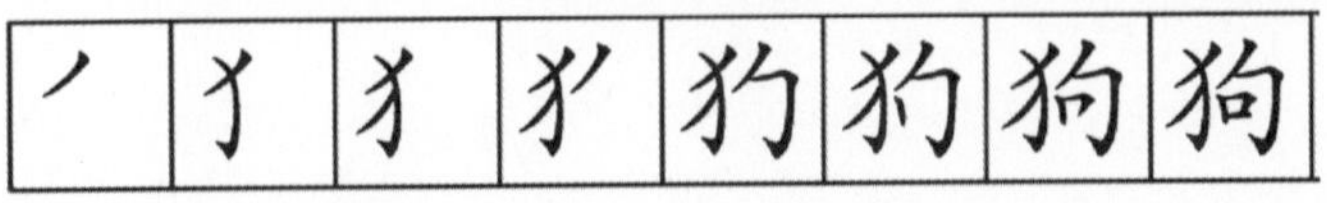

猫
māo

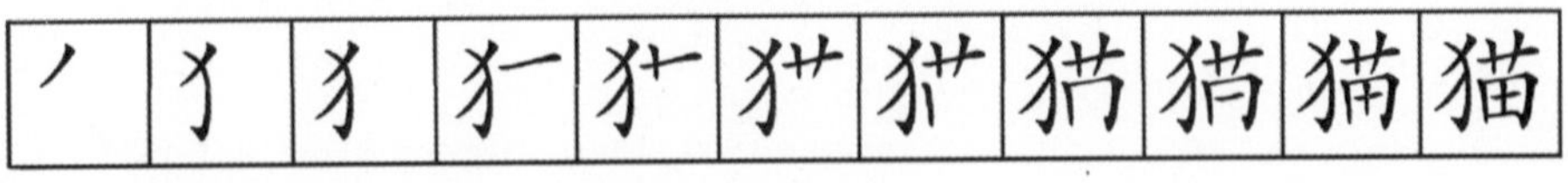

孙女
sūnnǚ

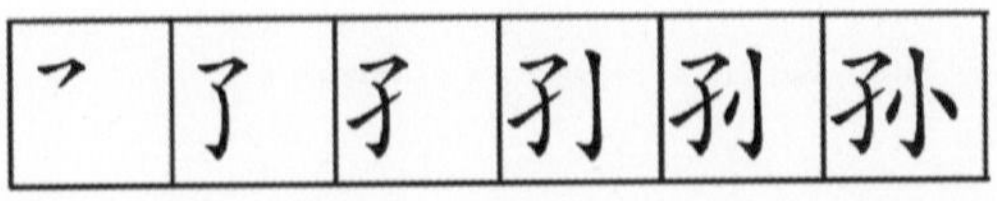

比较
bǐjiào

结婚
jiéhūn

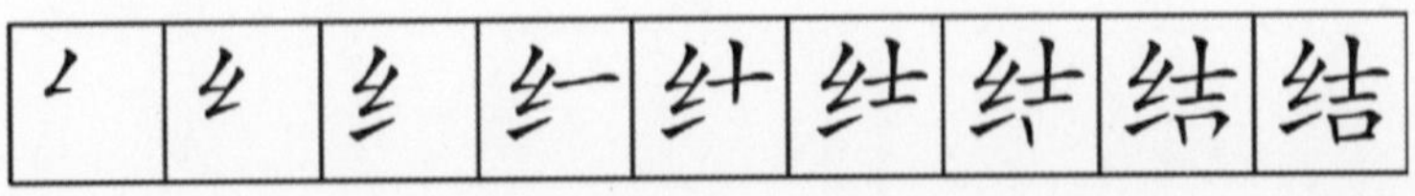

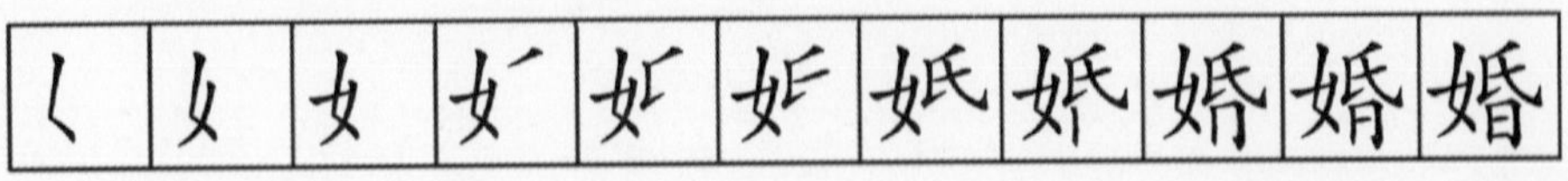

强
qiáng

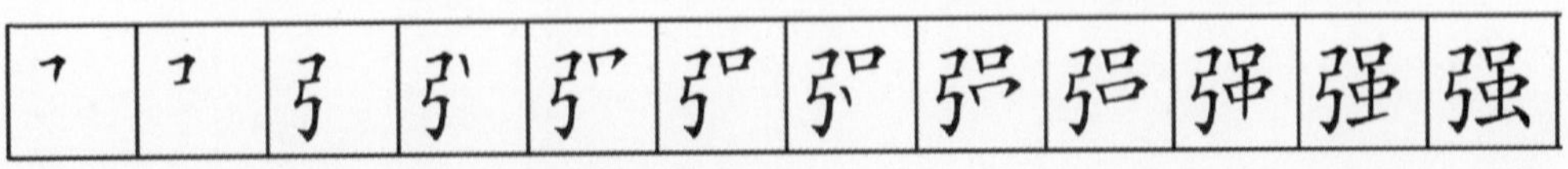

自己
zìjǐ

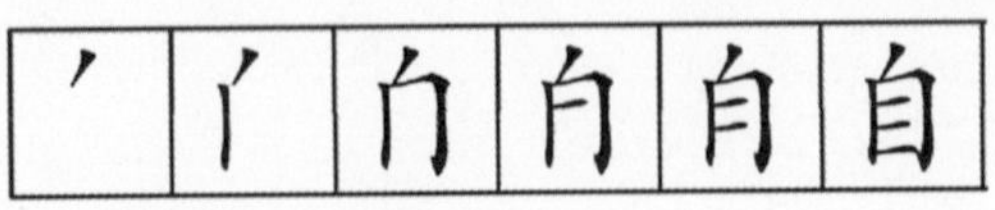

告诉
gàosu

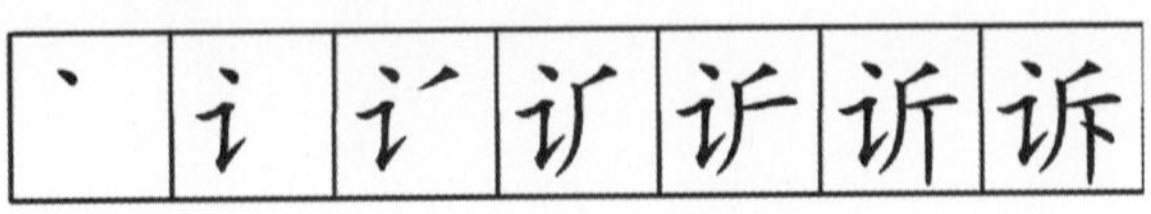

讨厌
tǎoyàn

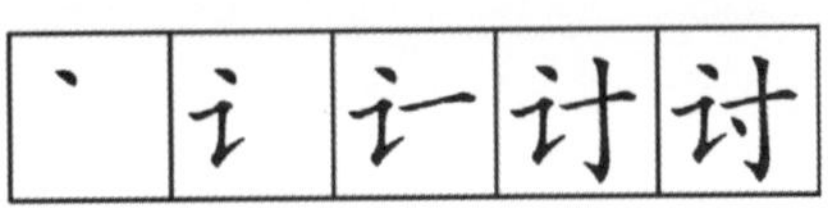

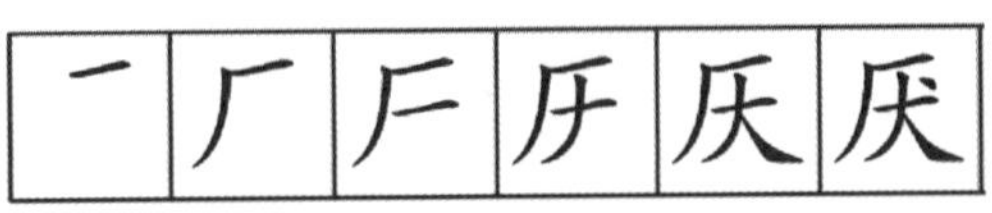

认识
rènshi

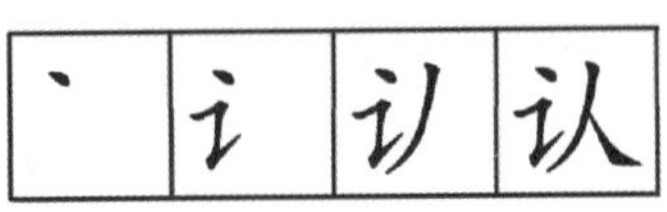

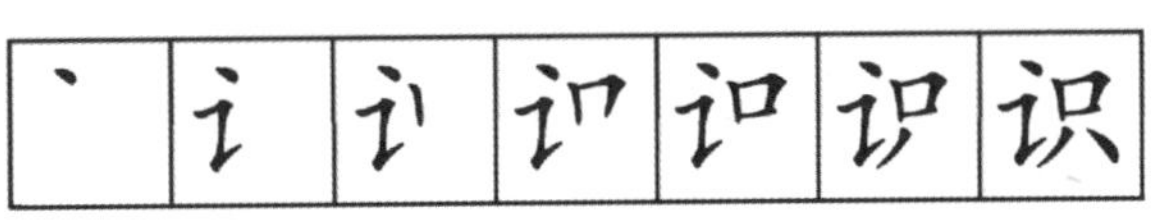

数学
shùxué

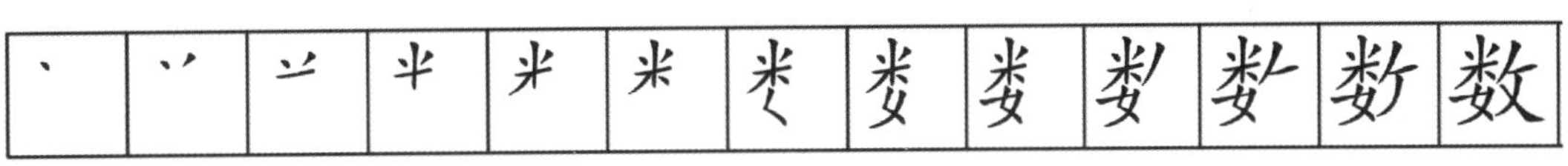

# Section 2 – Family life

Activity 16.14
Practise writing the following characters:

| | | | | | | | | | | | |
|---|---|---|---|---|---|---|---|---|---|---|---|
| 狗 | | | | | | | | | | | |
| 猫 | | | | | | | | | | | |
| 孙 | | | | | | | | | | | |
| 较 | | | | | | | | | | | |
| 结 | | | | | | | | | | | |
| 婚 | | | | | | | | | | | |
| 强 | | | | | | | | | | | |
| 诉 | | | | | | | | | | | |
| 讨 | | | | | | | | | | | |
| 厌 | | | | | | | | | | | |
| 认 | | | | | | | | | | | |
| 识 | | | | | | | | | | | |
| 自 | | | | | | | | | | | |
| 己 | | | | | | | | | | | |
| 数 | | | | | | | | | | | |

## Activity 16.15
Draw lines to match pairs and make logical phrases.

| | |
|---|---|
| 参加 | 结婚 |
| 告诉 | 婚礼 |
| 比较 | 号码 |
| 跟他 | 严肃 |
| 电话 | 孙女 |

## Activity 16.16
Split up the following words and write each under the correct heading according to the tone of each character. As an example, 比 has been completed for you.

比较　讨厌　婚礼　孙子　多大

| 1st tone | 2nd tone |
|---|---|
| | |
| **3rd tone** | **4th tone** |
| 比 | |

## Activity 16.17
Draw a line to match each Chinese word with its English translation.

| | |
|---|---|
| 饥荒 | to talk nonsense |
| 弱点 | to do something first |
| 强 | bad, broken |
| 先 | famine |
| 坏 | strong |
| 胡说 | cat |
| 猫 | weak point |

## Activity 16.18
Write the pinyin and tones of these sentences.

1. 我的炉子坏了。叔叔，我要买新的吗？
2. 别胡说！你到底觉得怎么样？
3. 他不清楚自己的弱点。
4. 你孙女滑冰的能力很强！
5. 很多地方都有饥荒的问题。

## Activity 16.19
Below, fill in the blanks with appropriate words.

1. 你的 ☐☐ 现在多大了？
2. 我的同事 ☐☐ 严肃。
3. 弟弟在 ☐☐ 里玩着手机。
4. 你的 ☐☐ 是哪年 ☐☐ 的？
5. 我 ☐ 去商场买东西，☐☐ 去爷爷家看他。

| | | |
|---|---|---|
| 父母 | 然后 | 比较 |
| 孙子 | 花园 | |
| 先 | 结婚 | |

## Activity 16.20

Translate the following sentences into Chinese.

1. I hate milk!

2. How old is your little sister?

3. I attended their wedding three months ago.

4. Please tell him I've gone to see the doctor.

5. My biology teacher is rather strict.

## Activity 16.21

Match the start and end of each sentence to make a logical complete sentence.

| | |
|---|---|
| 我真的不想 | 就给男朋友买吧。 |
| 如果你买得起， | 就飞到美国旅游。 |
| 你的冰箱没事儿， | 我和她不一样，我很讨厌经济学。 |
| 请不要告诉妈妈， | 现在可以用了。 |
| 我的孙子失业了以后， | 跟你结婚！ |

## Activity 16.22

Below, state how you feel about various things by finishing each sentence. Write in complete sentences (repeating the prompt text for extra practice).

1. 我比较喜欢……

2. 我很讨厌……

3. 我很怕……

4. 我对……很感兴趣。

### Activity 16.23

Read the passage about Asha and her granny's dog and answer the questions in English.

我很喜欢狗，可是我很讨厌奶奶的狗。它很小，可是它的眼睛和嘴都很大，狠不好看。我去看奶奶的时候，小狗一直在看着我。有一天我告诉奶奶说"你的狗让我感到不舒服，我很怕它！"可是奶奶只说"你别胡说！"然后对狗说"别听她的话，你是一只好狗狗。"

1. How does Asha feel about dogs?
2. How does she describe her granny's dog physically?
3. What does the dog do when she visits?
4. What did Asha tell her granny?
5. How did her granny respond?

### Activity 16.24

Answer the questions for yourself in Chinese. Use complete sentences.

1. 你有宠物吗？如果没有，你理想的宠物是什么？
2. 你的手机号码是多少？
3. 你觉得自己最强的能力是什么？
4. 你觉得自己最大的弱点是什么？

### Activity 16.25

Translate the following passage into English.

我的中国朋友上个月结婚了。他邀请了我去中国参加他的婚礼。我很高兴！ 婚礼很大，有几十个人来参加。我们都在大酒店一起吃饭。因为那时候我还不认识别的客人，所以我朋友让我跟他爸爸妈妈坐。客人都没有带礼物。大家都送红包。我朋友告诉我这是中国的习俗。

别的 biéde = other
习俗 xísú = customs

### Activity 16.26

Translate the following passage into Chinese.

My older brother's room is next to my room. I can hear everything he does. I really hate the music he puts on; I can hear it clearly. Although I told him I don't like it, he doesn't listen to me. He is only interested in himself. What do to?

# Unit 17 Section 1 Vocabulary

| | | |
|---|---|---|
| 附近 | fùjìn | nearby |
| 空气 | kōngqì | air |
| 天空 | tiānkōng | sky |
| 灰色 | huīsè | grey |
| 新鲜 | xīnxiān | fresh |
| 钓鱼 | diào yú | to fish |
| 污染 | wūrǎn | pollution |
| 公里 | gōnglǐ | kilometre |
| 不但 | búdàn | not only |
| 而且 | érqiě | furthermore |
| 浪费 | làngfèi | to waste |
| 志愿者 | zhìyuànzhě | volunteer (worker) |
| 协会 | xiéhuì | an association |
| 自然 | zìrán | nature, natural |
| 河 | hé | river |
| 垃圾 | lājī | rubbish |
| 种 | zhòng | to plant |
| 树 | shù | tree |
| 山区 | shānqū | mountain area |
| 气候变化 | qìhòu biànhuà | climate change |
| 比如 | bǐrú | for example |
| 等等 | děngděng | et cetera |
| 政府 | zhèngfǔ | government |
| 战争 | zhànzhēng | war |

**Online vocabulary**

| | | |
|---|---|---|
| 熊猫 | xióngmāo | panda |
| 暑假 | shǔjià | summer holidays |
| 寒假 | hánjià | winter holidays |
| 星星 | xīngxīng | star |

# Unit 17 Section 1 Stroke Order

新鲜
xīnxiān

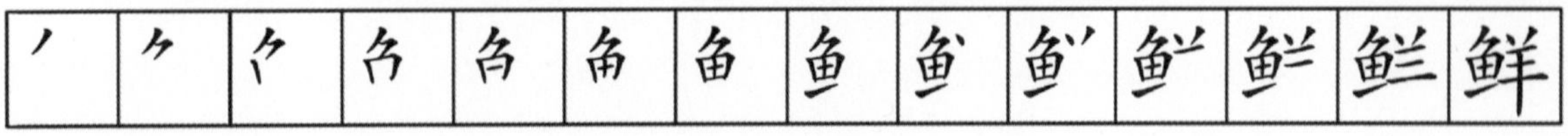

钓鱼
diào yú

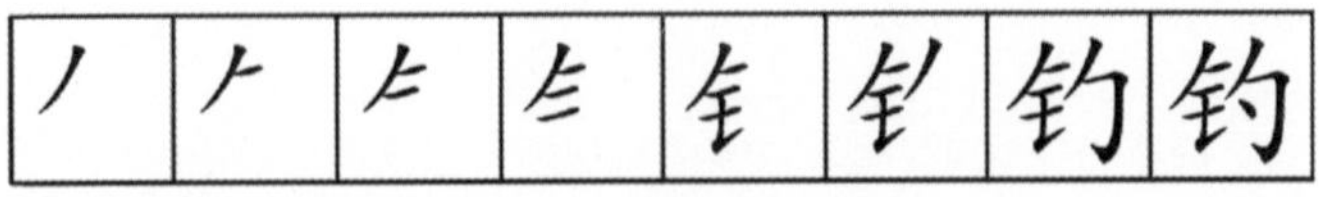

污染
wūrǎn

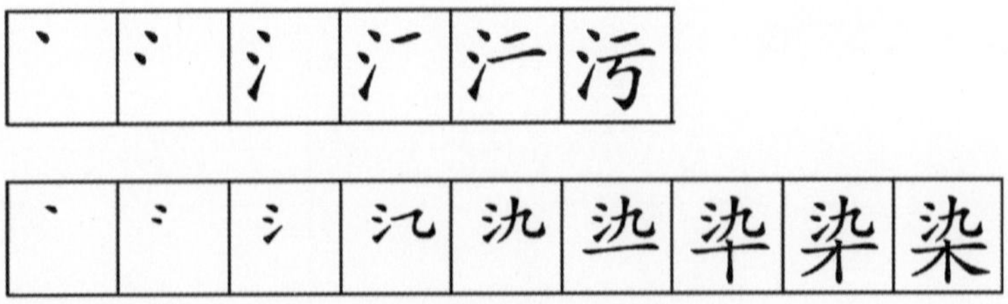

而且
érqiě

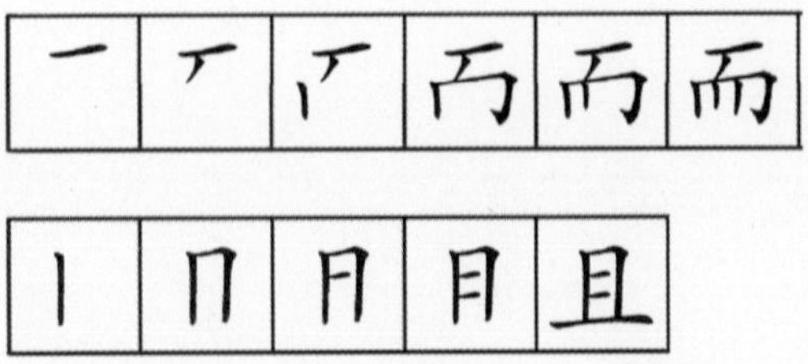

浪费
làngfèi

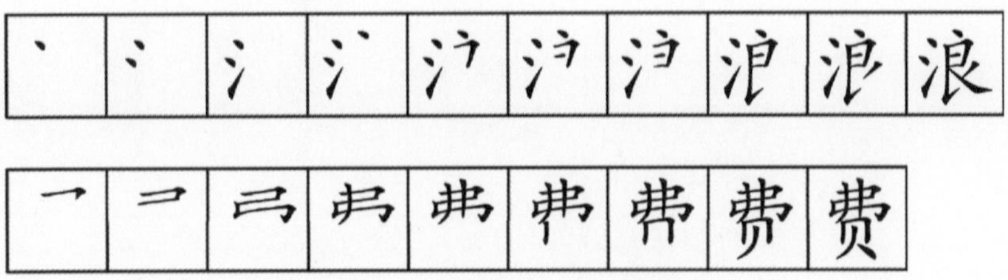

志愿者
zhìyuànzhě

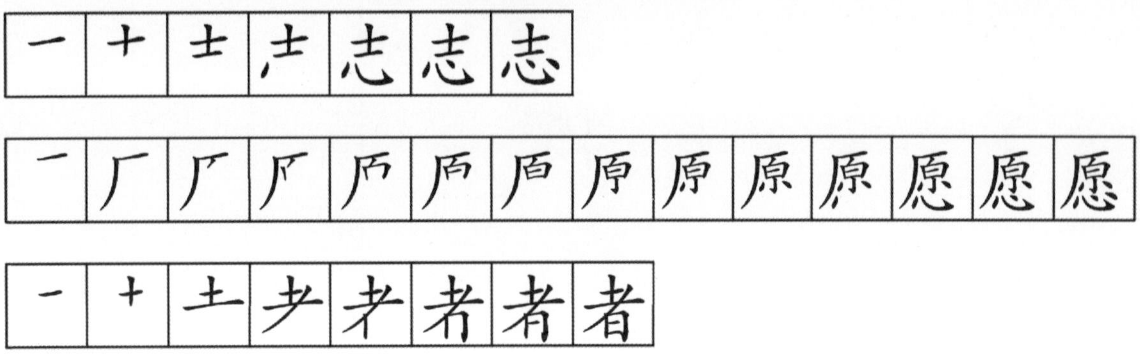

河
hé

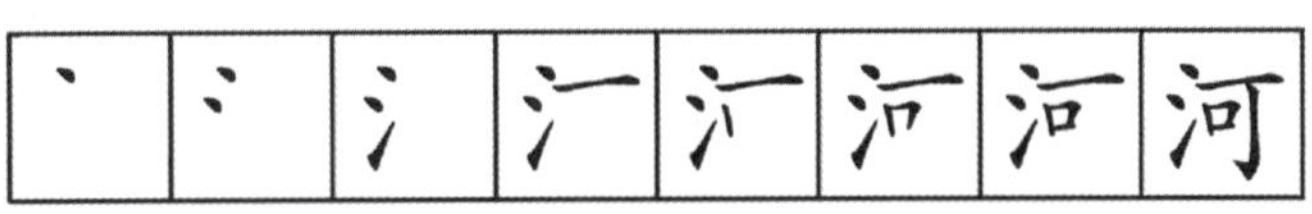

种
zhòng

树
shù

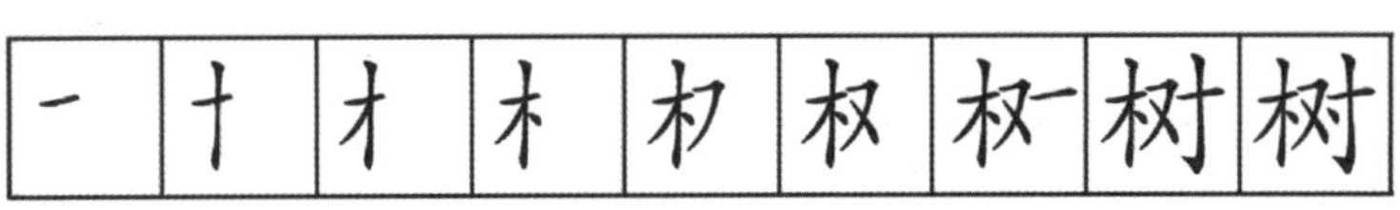

# 全球问题 Global Issues

## Section 1 – Environment

Activity 17.1

Practise writing the following characters:

鲜

钓

污

染

而

且

浪

费

志

愿

者

河

种

树

### Activity 17.2
Draw lines to match pairs and make logical phrases.

| | |
|---|---|
| 空气 | 树 |
| 浪费 | 污染 |
| 十八 | 寒假 |
| 种 | 公里 |
| 放 | 时间 |

### Activity 17.3
Split up the following words and write each under the correct heading according to the tone of each character. As an example, 空 has been completed for you.

空气　自然　新鲜　钓鱼　浪费

| 1st tone | 2nd tone |
|---|---|
| 空 | |
| **3rd tone** | **4th tone** |
| | |

### Activity 17.4
Draw a line to match each Chinese word with its English translation.

| | |
|---|---|
| 协会 | for example |
| 比如 | panda |
| 战争 | an association |
| 熊猫 | government |
| 灰色 | war |
| 垃圾 | rubbish |
| 政府 | grey |

### Activity 17.5
Write the pinyin and tones of these sentences.

1. 你看到过中国的网站吗?
2. 这样不但能省钱，而且能省时间。
3. 很糟糕，哪儿都有垃圾。
4. 我最担心气候变化的问题。
5. 你对职业足球协会感兴趣吗?

### Activity 17.6
Below, fill in the blanks with appropriate words.

| | | |
|---|---|---|
| 志愿者 | 大自然 | 浪费 |
| 钓鱼 | 风景 | 空气 |
| 河 | 种树 | |

1. 上个月我当□□□去公园□□了。
2. 他喜欢在□□□中散步。
3. 那位老师从来不□□纸。
4. 他住在□附近，每天都去□□。
5. 这个地方真好：□□又很漂亮，□□又很新鲜。

## Activity 17.7

Translate the following sentences into Chinese.

1. The air is so fresh!

2. Pollution can affect everyone's health.

3. Before, many people in my city went fishing. Now the fish are very few.

4. The place I live has many problems: for example, pollution, crime, etc.

5. I not only want to save more water, but also be a "Green Volunteer".

## Activity 17.8

Match the start and end of each sentence to make a logical complete sentence.

| | |
|---|---|
| 我的小镇有多穷人， | 他很担心气候变化。 |
| 河边以前很脏， | 对熊猫感兴趣。 |
| 社会问题很多， | 可是小偷很少。 |
| 哥哥每天看新闻， | 现在很干净了。 |
| 最近越来越多人 | 可是做志愿者的人很少。 |

## Activity 17.9

Translate the sentences below into Chinese without using 比.

1. Today I feel a bit happier.

2. This way, it's a little clearer.

3. Your house is a lot dirtier!

4. I want to speak a little more fluently.

5. This resume is much better!

### Activity 17.10
Read the passage about Samantha's hobby and answer the questions in English.

我喜欢晚上去外边看星星。只有天气不太冷，我每天晚上都去。上个学期我出国旅游了。我和同学在大城市住了三天。第一天晚上我去河边看了天空，可是因为空气污染，看不见星星。那时候，天空是灰色的。小河也很脏。我觉得很不高兴。第二天没有污染，我能看到星星了。

严重 yánzhòng = serious
一会儿 yī huìr = a little while

1. What is Samantha's hobby?
2. When does she not go out to do this activity?
3. When did she go travelling?
4. What was the problem on her first evening there?
5. How does she describe the river?

### Activity 17.11
Answer the questions for yourself in Chinese. Use complete sentences.

1. 你住的地方有些什么环境问题？
2. 哪些环境问题最重要？
3. 你做过志愿者吗？做了什么事儿？
4. 你觉得钓鱼怎么样？

### Activity 17.12
Translate the following passage into English.

这个地方的社会问题不少。我们这儿有毒品问题、住房问题和交通问题。环境污染也越来越严重。有水污染、空气污染、等等。怎么办？有时候我想去山区生活。山区的生活不但很舒服，而且很健康。

严重 yánzhòng = serious

### Activity 17.13
Translate the following passage into Chinese.

My dad loves to fish, but I don't. He will walk several kilometres to get to the river. I think it is a waste of time. There are more important things to do, like volunteer work, sending emails, reading, etc. Dad doesn't understand. He says that everyone is too busy and that fishing in a quiet place is beneficial to one's health.

# Unit 17 Section 2 Vocabulary

| | | |
|---|---|---|
| 网吧 | wǎngba | internet café |
| 上瘾 | shàngyǐn | to be addicted to |
| 酒 | jiǔ | alcohol |
| 吸烟 | xī yān | to smoke |
| 戒烟 | jiè yān | to quit smoking |
| 并 | bìng | actually (not) |
| 饮食 | yǐnshí | food and drink |
| 超重 | chāozhòng | overweight |
| 减肥 | jiǎn féi | to lose weight |
| 因此 | yīncǐ | consequently |
| 洪水 | hóngshuǐ | flood |
| 人口 | rénkǒu | population |
| 增加 | zēngjiā | to increase |
| 同意 | tóngyì | to agree |
| 反对 | fǎnduì | to oppose |

**Online vocabulary**

| | | |
|---|---|---|
| 吵 | chǎo | noisy |
| 热闹 | rènào | bustling with activity |

# Unit 17 Section 2 Stroke Order

山区
shānqū

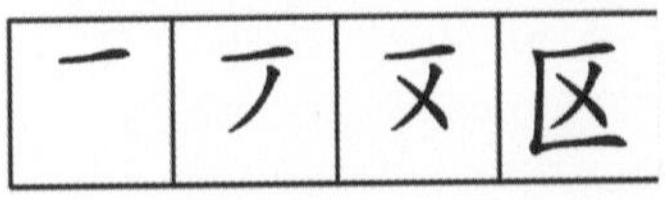

担心
dānxīn

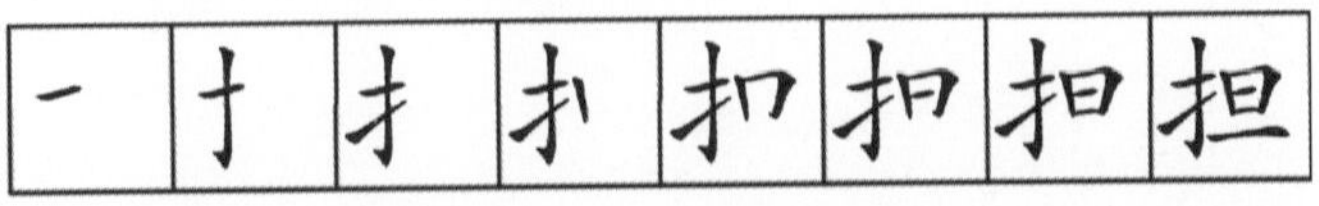

变化
biànhuà

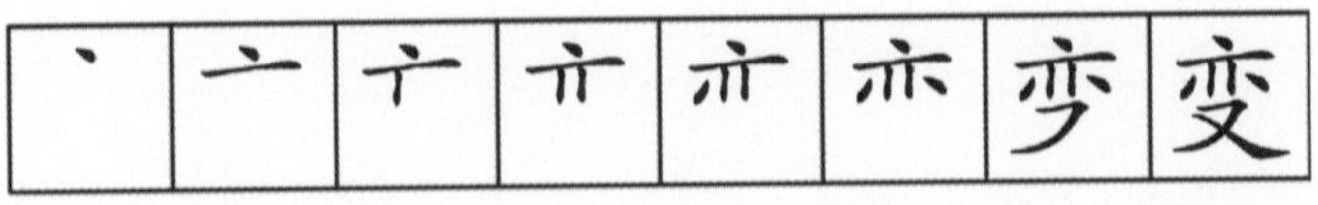

熊猫
xióngmāo

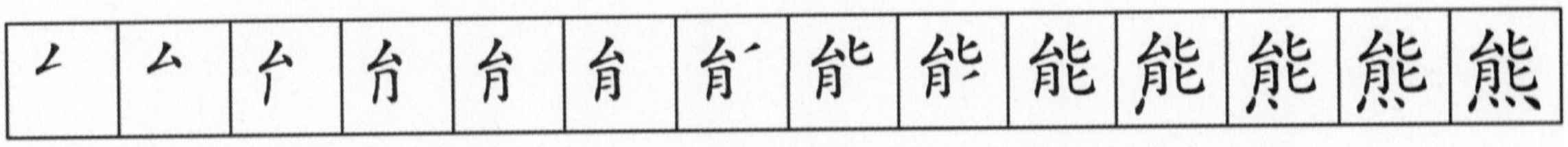

暑假
shǔjià

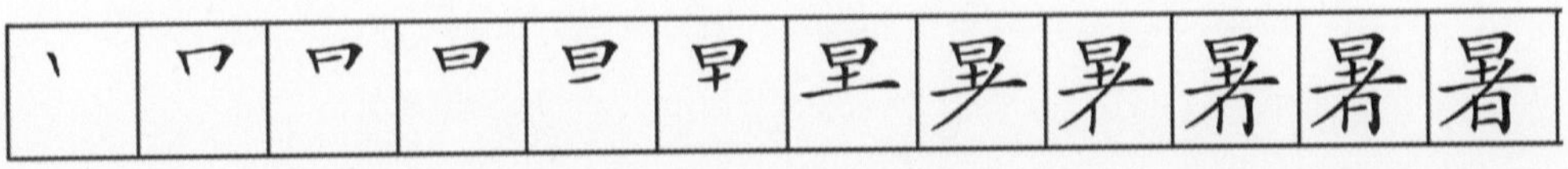

寒假
hánjià

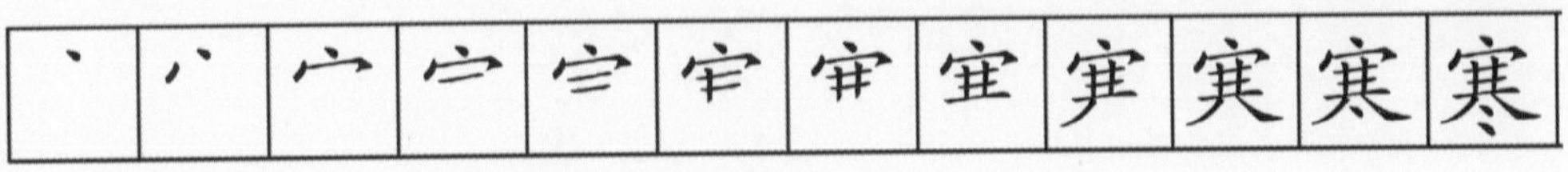

吸烟
xī yān

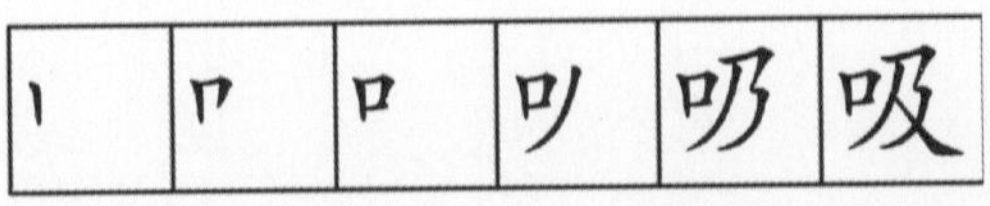

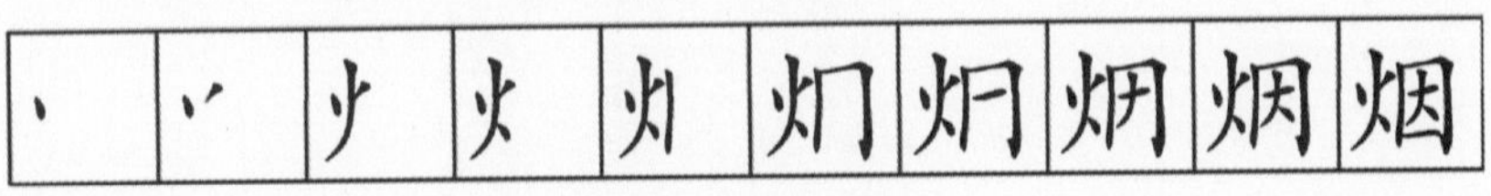

并
bìng

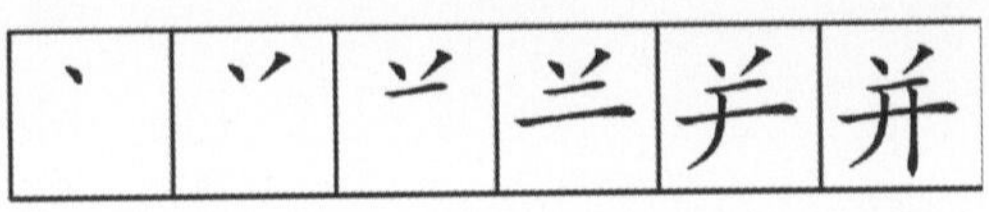

饮食
yǐnshí

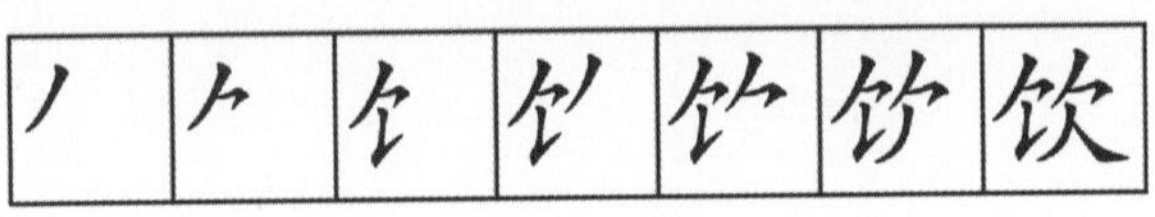

因此
yīncǐ

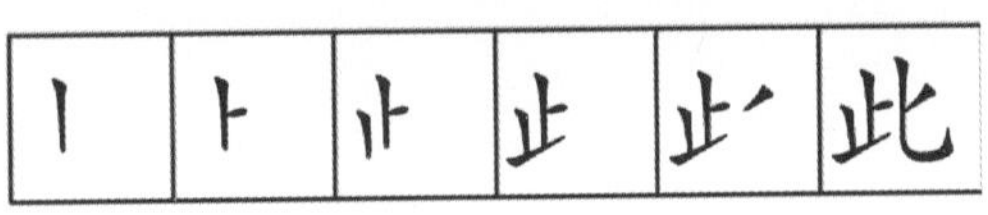

反对
fǎnduì

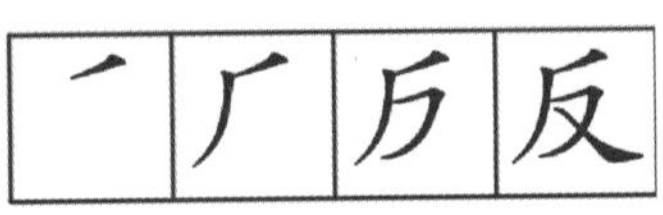

吵
chǎo

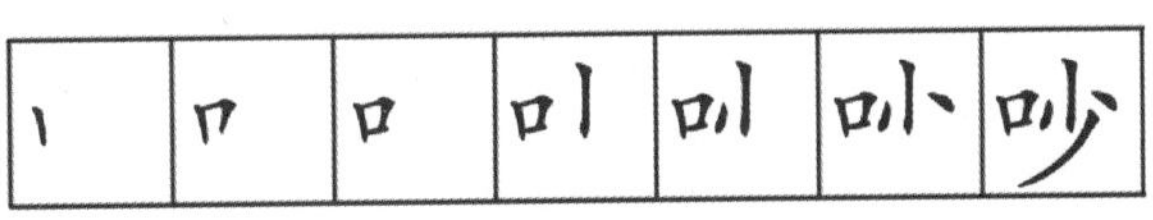

热闹
rènào

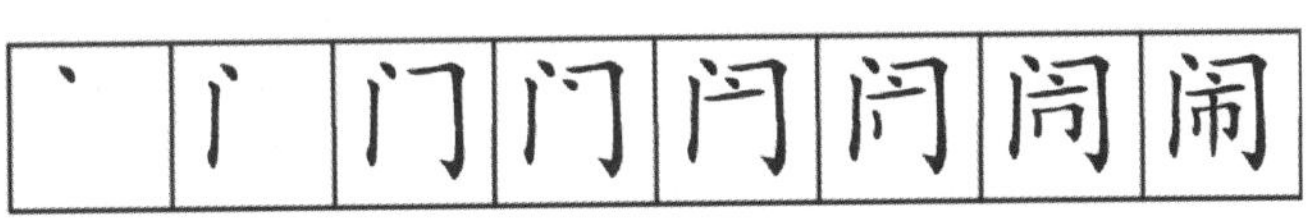

# Section 2 – Social issues

Activity 17.14
Practise writing the following characters:

| | | | | | | | | | | | |
|---|---|---|---|---|---|---|---|---|---|---|---|
| 区 | | | | | | | | | | | |
| 担 | | | | | | | | | | | |
| 变 | | | | | | | | | | | |
| 熊 | | | | | | | | | | | |
| 暑 | | | | | | | | | | | |
| 寒 | | | | | | | | | | | |
| 吸 | | | | | | | | | | | |
| 烟 | | | | | | | | | | | |
| 并 | | | | | | | | | | | |
| 饮 | | | | | | | | | | | |
| 此 | | | | | | | | | | | |
| 反 | | | | | | | | | | | |
| 吵 | | | | | | | | | | | |
| 闹 | | | | | | | | | | | |

## Activity 17.15
Draw lines to match pairs and make logical phrases.

| | |
|---|---|
| 人口 | 习惯 |
| 饮食 | 活动 |
| 戒 | 烟 |
| 附近的 | 增加 |
| 参加 | 网吧 |

## Activity 17.16
Split up the following words and write each under the correct heading according to the tone of each character. As an example, 因 has been completed for you.

因此　　人口　　热闹　　吸烟　　饮食

| 1st tone | 2nd tone |
|---|---|
| 因 | |
| **3rd tone** | **4th tone** |
| | |

## Activity 17.17
Draw a line to match each Chinese word with its English translation.

| | |
|---|---|
| 超重 | noisy |
| 洪水 | overweight |
| 反对 | to oppose |
| 减肥 | flood |
| 上瘾 | to be addicted |
| 吵 | to quit smoking |
| 戒烟 | to lose weight |

## Activity 17.18
Write the pinyin and tones of these sentences.

1. 他对酒早就上瘾了，可是他说他没有问题。

2. 她认为少吃东西可以减肥，可是我反对这个想法。

3. 我担心儿子超重了，可是我不知道怎么办。

4. 谁都知道，吸烟对身体有害。

5. 四月是洪水季节。

## Activity 17.19
Below, fill in the blanks with appropriate words.

1. 你父亲☐☐对你的身体有☐☐吗？

2. 他觉得☐☐超重了，可是我并不☐☐。

3. 你喜欢很☐☐的地方吗？

4. 阿姨每天都去☐☐上网。

| | | |
|---|---|---|
| 同意 | 网吧 | 吸烟 |
| 自己 | 影响 | 热闹 |

## Activity 17.20

Translate the following sentences into Chinese.

1. I am most concerned about climate change.

2. He is addicted to his mobile phone.

3. I agree, this place is too noisy.

4. The internet café is 9 kilometres from here. Too far, we can't walk (there).

5. He wastes time drinking after getting out of work each day.

## Activity 17.21

Match the start and end of each sentence to make a logical complete sentence.

| | |
|---|---|
| 如果你要减肥， | 我要减肥。 |
| 她不但戒了烟， | 而且不喝酒了。 |
| 爷爷反对孙子去网吧， | 洪水不是常见的问题。 |
| 在山区， | 就可以去健身房做运动。 |
| 我超重了， | 他担心孩子会不安全。 |

## Activity 17.22

Below, write a sentence for each issue suggesting a solution.

1. 你朋友想跟你放鞭炮，可是你不喜欢这个活动。

2. 一个同学从来不关灯。

3. 坐公共汽车时，旁边的人开始吸烟。

4. 图书馆里很吵，你不能做作业。

## Activity 17.23

Read the passage about Mario's summer holiday and answer the questions in English.

我们六月二十号放暑假。我和弟弟去自然保护区做了义工。一个月以前，那儿附近有洪水，因此我们跟学校的志愿者协会去帮助当地人。我从来没去过山区，没想到空气会那么干净！我们住的小镇也非常安静。有人说那个小镇的人口越来越少，因为没有工作机会。

当地人 dāngdìrén = local person

1. When did Mario start summer break?
2. Who arranged the trip to the mountain village?
3. Why were they there?
4. What did Mario discover?
5. Give a reason for the social trend he mentions.

## Activity 17.24

Answer the questions for yourself in Chinese. Use complete sentences.

1. 有人说吃肉对身体不好。你同意吗？
2. 有人说喝点儿酒对身体有害。你同意吗？
3. 你放暑假的时候喜欢做什么？
4. 你放寒假的时候喜欢做什么？

## Activity 17.25

Translate the following passage into English.

妈妈说我放寒假的时候，她带我去滑雪。我应该很高兴，可是我并不想去。第一，滑雪太贵了。第二，滑雪很不安全，很容易受伤。第三，滑雪场浪费电，对环境不好。我跟妈妈说，我要找更好的寒假活动。

## Activity 17.26

Translate the following passage into Chinese.

My dad smokes often. He started smoking when I was young. I really hate smoking, but whatever I say, he won't listen. He not only gets angry, but also says he doesn't have a problem at all. What can I do?

# Unit 18 Section 1 Vocabulary

| | | |
|---|---|---|
| 桔子 | júzi | tangerine |
| 公斤 | gōngjīn | kilogram |
| 水果 | shuǐguǒ | fruit |
| 蔬菜 | shūcài | vegetables |
| 汽水 | qìshuǐ | soda |
| 瓶 | píng | bottle |
| 入口 | rùkǒu | entrance |
| 出口 | chūkǒu | exit |
| 当然 | dāngrán | of course |
| 节目 | jiémù | programme |
| 名人 | míngrén | celebrity |
| 海鲜 | hǎixiān | seafood |
| 展览 | zhǎnlǎn | to exhibit, exhibition |
| 购物 | gòu wù | to go shopping |
| 咖啡厅 | kāfēitīng | café |
| 菜单 | càidān | menu |
| 烧 | shāo | to burn, braise, stew |

**Online vocabulary**

| | | |
|---|---|---|
| 盘 | pán | plate |
| 满意 | mǎnyì | to be satisfied |
| 快餐 | kuàicān | fast food |
| 青菜 | qīngcài | green vegetables, Chinese cabbage |
| 方便 | fāngbiàn | convenient |

# Unit 18 Section 1 Stroke Order

| Word | Stroke order |
| --- | --- |
| 公斤 gōngjīn | 丿 厂 斤 斤 |
| 入口 rùkǒu | 丿 入 |
| 购物 gòu wù | 丨 冂 贝 贝 贝 购 购 购 |
| 咖啡 kāfēi | 丨 冂 口 叻 叻 叻 咖 咖 |
| | 丨 冂 口 口 口 口 口 啡 啡 啡 啡 |
| 盘 pán | 丿 丿 ⺆ 舟 舟 舟 舟 盘 盘 盘 盘 |
| 满意 mǎnyì | 丶 ⺀ 氵 汁 汁 汁 泄 泄 满 满 满 满 满 |
| 青菜 qīngcài | 一 二 丰 主 青 青 青 青 |
| 方便 fāngbiàn | 丿 亻 亻 亻 佰 佰 佰 便 便 |
| 孩子 háizi | 乛 了 子 孑 孩 孩 孩 孩 孩 |
| 篮球 lánqiú | 丿 𠂉 𥫗 𥫗 ⺮ ⺮ 笁 笁 笁 笁 笁 笁 篮 篮 篮 篮 |
| 累 lèi | 丨 冂 冂 田 田 田 罒 累 累 累 累 |
| 纸 zhǐ | 𠃋 纟 纟 纟 纸 纸 纸 |
| 苹果 píngguǒ | 一 十 艹 艹 艹 艹 苹 苹 |

# 出去玩儿 Going Out

## Section 1 – Food and cuisine

Activity 18.1

Practise writing the following characters:

斤

入

购

咖

啡

盘

满

青

便

孩

篮

累

纸

苹

### Activity 18.2
Draw lines to match words with their simple definitions.

| | |
|---|---|
| 快餐 | 比如苹果 |
| 水果 | 比如虾 |
| 蔬菜 | 帮你点菜的一张纸 |
| 海鲜 | 马上准备好的食物 |
| 菜单 | 比如青菜 |

### Activity 18.3
Split up the following words and write each under the correct heading according to the tone of each character. As an example, 公 has been completed for you.

公斤　购物　满意　当然　节目

| 1st tone | 2nd tone |
|---|---|
| 公 | |
| **3rd tone** | **4th tone** |
| | |

### Activity 18.4
Draw a line to match each Chinese word with its English translation.

| | |
|---|---|
| 瓶 | child |
| 烧 | bottle |
| 盘 | tangerine |
| 桔子 | menu |
| 孩子 | to braise |
| 菜单 | to exhibit, exhibition |
| 展览 | plate |

### Activity 18.5
Write the pinyin and tones of these sentences.

1. 这个孩子讨厌蔬菜和水果。
2. 那边是入口，这边是出口。
3. 你昨天去看的展览怎么样？
4. 来两盘红烧牛肉。
5. 这个名人影响到了年轻人。

### Activity 18.6
Below, fill in the blanks with appropriate words.

1. 我听说很多作家喜欢带电脑去☐☐☐工作。
2. 他吃午饭的时候，总喝一瓶☐☐。
3. 她最近对☐☐上瘾了！
4. 我觉得非常☐☐！
5. 我想买☐☐，请给我三☐☐左右。

| | |
|---|---|
| 满意 | 苹果 |
| 汽水 | 海鲜 |
| 公斤 | 图书馆 |

## Activity 18.7

Translate the following sentences into Chinese.

1. You can buy tickets over there. See the entrance?

2. Of course you can come shopping with us!

3. We were very satisfied with the dumplings, so we went back to the restaurant

4. Taking the bus is very convenient. I'm not at all tired.

5. Will you attend the basketball match? I heard the athletes are all excellent!

## Activity 18.8

Match the start and end of each sentence to make a logical complete sentence.

| | |
|---|---|
| 坐一会儿吧， | 对身体有害。 |
| 这个电视节目的演员 | 我叫服务员过来 。 |
| 妈妈经常说快餐 | 我们常常给朋友送桔子。 |
| 过年的时候， | 你不要走！ |
| 请看一下菜单， | 都不是名人。 |

## Activity 18.9

Below are a few items of food and drink. Write a sentence stating how you feel about each.

1. 苹果

2. 青菜

3. 咖啡

4. 海鲜

5. 汽水

## Activity 18.10

Read the passage about Raquel's outing and answer the questions in English.

我朋友对化学很感兴趣。因为上个星期是她的生日，所以我带她去了附近博物馆的科学展览。展览里有很多好玩活动。有很多孩子看展览，所以里面很吵。看完展览以后，我们并不累，打算去看电影。那时候我们还没吃饭，没有很多时间，所以我们找了一家快餐店。虽然快餐不太好吃，但是它又便宜又方便。

1. Why did Raquel take her friend to an exhibition?
2. Where was the exhibition held?
3. What was at the exhibition?
4. Why did they get fast food afterwards?
5. What comment does Raquel make about the food?

## Activity 18.11

Answer the questions for yourself in Chinese. Use complete sentences.

1. 你最喜欢的水果是什么？
2. 你最喜欢的蔬菜是什么？
3. 你最喜欢吃哪国的菜？
4. 你吃过成都菜吗？
5. 你觉得快餐的好处是什么？坏处呢？

## Activity 18.12

Translate the following passage into English.

我的儿子以前很喜欢吃海鲜和鱼。一天，他看到了电视节目说海洋里的鱼越来越少了。然后，儿子不想吃鱼了。他还让我看这个节目。我同意他的想法。这个电视节目对我们的生活有很大的影响！

## Activity 18.13

Translate the following passage into Chinese.

My older sister is a celebrity. Because her work is always so busy, she isn't very healthy. She drinks too much coffee and eats too much fast food. She says this way is convenient. She also never exercises and doesn't sleep well. I am worried about her.

# Unit 18 Section 2 Vocabulary

| | | |
|---|---|---|
| 准备 | zhǔnbèi | to prepare |
| 美术馆 | měishùguǎn | art gallery |
| 一边……，一边…… | yībiān...yībiān... | doing (sth) while doing (something else) |
| 乐趣 | lèqù | delight, pleasure |
| 照相 | zhào xiàng | to take a photo |
| 滑板 | huábǎn | skateboard |
| 音乐会 | yīnyuè huì | concert |
| 零用钱 | língyòngqián | pocket money |
| 够 | gòu | to be enough |
| 成绩 | chéngjī | achievement, grades |
| 讲 | jiǎng | to speak |
| 可能 | kěnéng | perhaps |
| 独生子 | dúshēngzǐ | only child |

**Online vocabulary**

| | | |
|---|---|---|
| 行 | xíng | okay, acceptable |
| 记得 | jìdé | to remember |
| 必须 | bìxū | must |
| 不好意思 | bùhǎo yìsi | embarrassed, "Excuse me" |
| 一定 | yīdìng | certainly |

# Unit 18 Section 2 Stroke Order

准备
zhǔnbèi

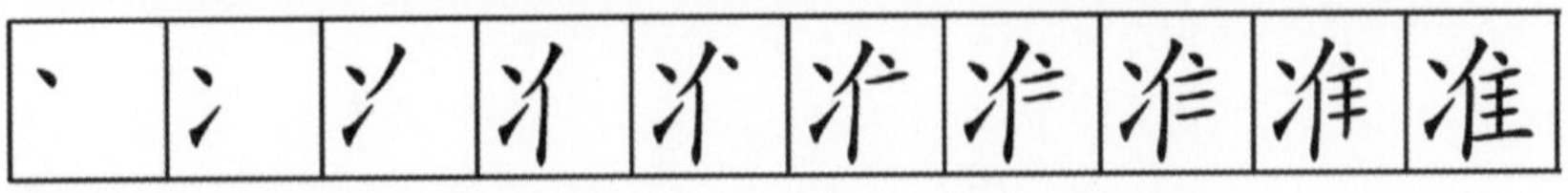

照相
zhàoxiàng

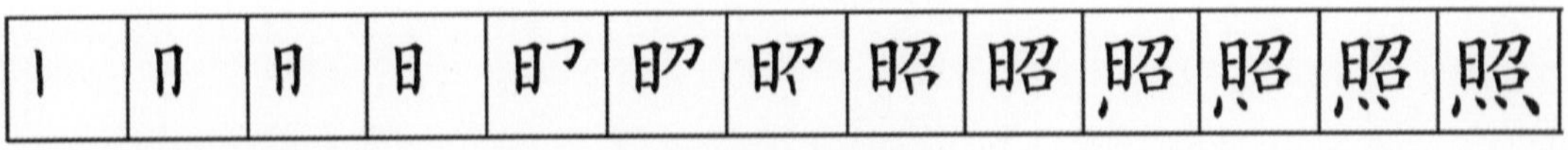

滑板
huábǎn

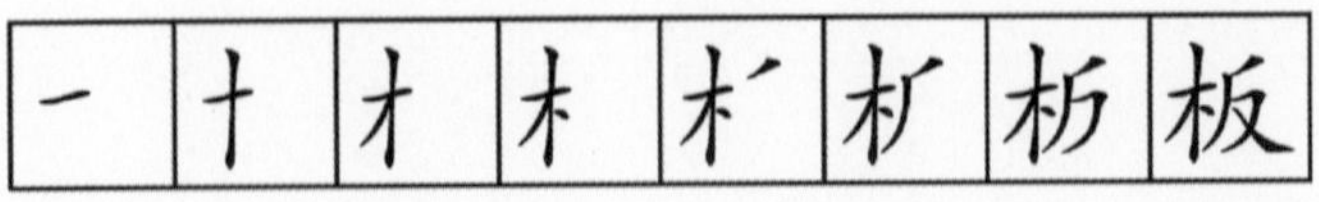

零用钱
língyòng qián

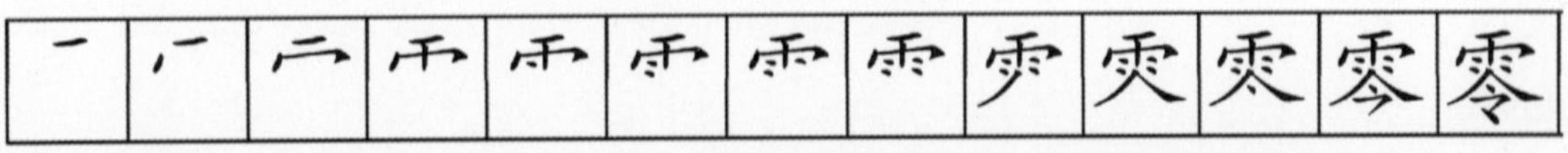

够
gòu

成绩
chéngjī

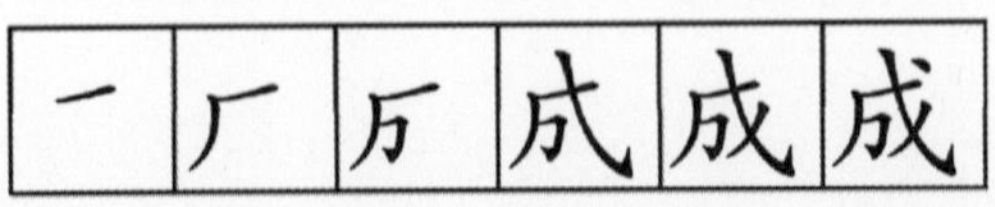

讲
jiǎng

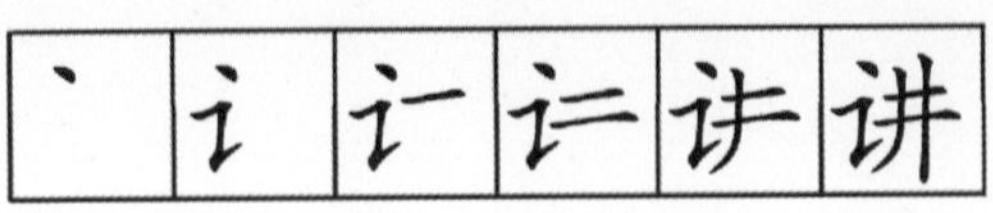

独生子
dúshēng zǐ

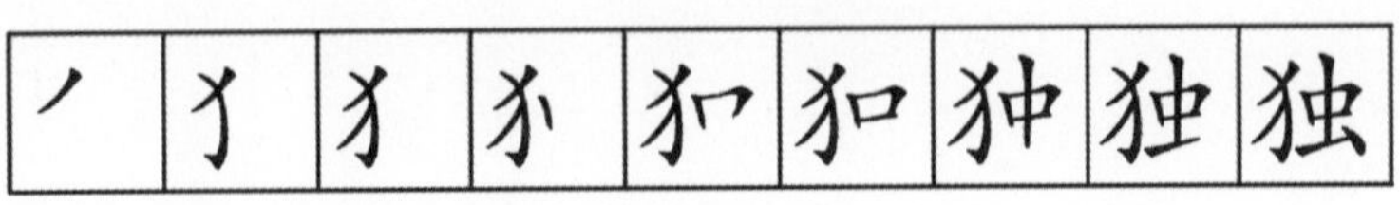

记得
jìdé

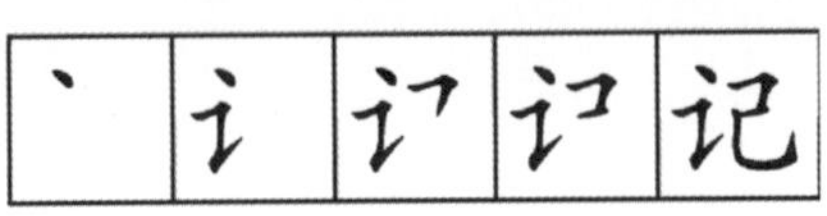

必须
bìxū

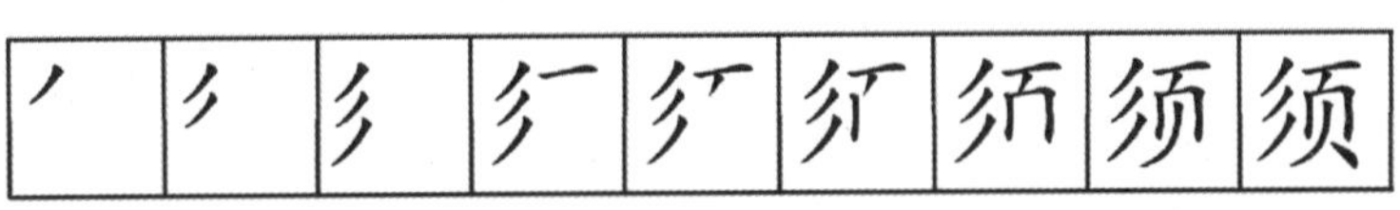

一定
yīdìng

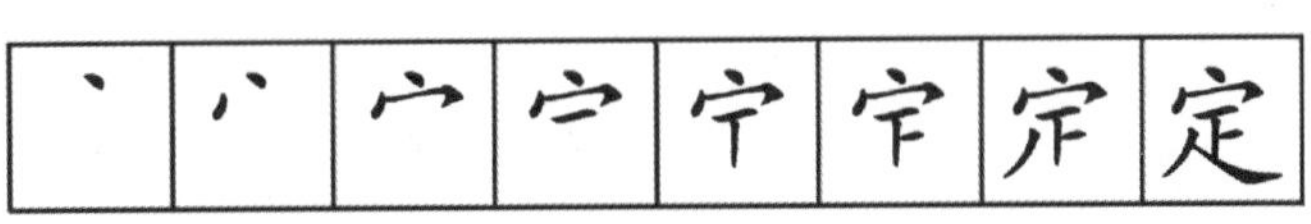

外
wài

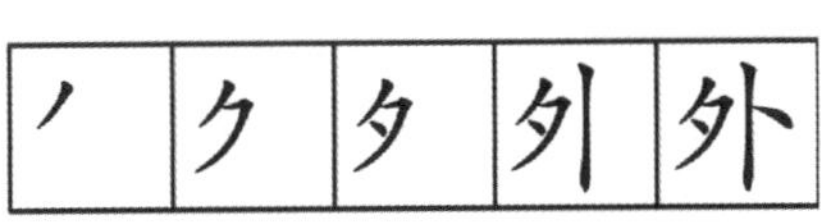

## Section 2 – Free time

Activity 18.14
Practise writing the following characters:

准
备
照
板
零
够
成
绩
讲
独
记
须
定
外

## Activity 18.15
Draw lines to match pairs and make logical phrases.

| | |
|---|---|
| 记得 | 小时候 |
| 必须 | 旅行袋 |
| 对成绩 | 零用钱 |
| 准备 | 满意 |
| 花 | 遛狗 |

## Activity 18.16
Split up the following words and write each under the correct heading according to the tone of each character. As an example, 必 has been completed for you.

必须　乐趣　滑板　准备　照相

| 1st tone | 2nd tone |
|---|---|
| | |
| **3rd tone** | **4th tone** |
| | 必 |

## Activity 18.17
Draw a line to match each Chinese word with its English translation.

| | |
|---|---|
| 成绩 | okay, acceptable |
| 可能 | achievement, grades |
| 不好意思 | to remember |
| 行 | perhaps |
| 乐趣 | to speak, narrate |
| 讲 | delight, pleasure |
| 记得 | embarrassed, "Excuse me" |

## Activity 18.18
Write the pinyin and tones of these sentences.

1. 你打算和他一起参加音乐会吗?
2. 孩子没有零用钱怎么能去购物呢?
3. 旅游的时候，你一定要拍照片!
4. 父母不让我滑滑板，因为不安全。
5. 蔬菜我们已经点够了，不用再点了。

## Activity 19.19
Below, fill in the blanks with appropriate words.

1. 爸爸对我的 ☐☐ 很不 ☐☐ 。
2. 我先 ☐☐ 晚饭，然后和你玩儿 ☐☐ ，好吗?
3. 生活的 ☐☐ 怎么能找到?
4. 我早就 ☐☐ 你的意思。

满意　游戏　乐趣　明白　准备　成绩

## Activity 18.20

Translate the following sentences into Chinese.

1. Was your grandmother an only child?

2. He's in the bedroom, talking on the phone and playing with the cat (at the same time).

3. Sorry, I can't go the concert, I've become ill.

4. My mother is very satisfied with my exam score.

5. I want to visit the art gallery, but there's not enough time.

## Activity 18.21

Match the start and end of each sentence to make a logical complete sentence.

| | |
|---|---|
| 不好意思, | 你要多穿衣服! |
| 他跑了很长时间, | 你可能会觉得不舒服 。 |
| 天气变冷了, | 他一定很累! |
| 为了将来的工作, | 这次考试很重要。 |
| 如果你没坐过船, | 我不认识您。您贵姓? |

## Activity 18.22

Below are various places of leisure. Write a sentence for each stating how many times you go per year (or week/ month, as appropriate).

1. 音乐会
2. 美术馆
3. 历史博物馆
4. 海滩
5. 体育馆
6. 网吧
7. 森林
8. 书店

## Activity 18.23

Read the passage about how Zhang Bo spends his free time and answer the questions in English.

我每天都去健身房。这样不但很健康，而且很有趣。我的学校有健身房，但是很小，也没有电视机。我常常去城市的公共体育馆，因为那里有电视机。我喜欢一边健身一边看电视。我最喜欢的电视节目是运动节目和台湾的电视剧。周末，我跟朋友去市中心购物。虽然我没有钱，可是我喜欢看看商店的商品。

1. Why doesn't Zhang Bo use the gym at his school?
2. What does Zhang Bo do while working out?
3. What is Zhang Bo's favourite type of TV show?
4. Where does Zhang Bo go at the weekend?
5. What does he do there?

## Activity 18.24

Answer the questions for yourself in Chinese. Use complete sentences.

1. 你去旅游以前会做什么准备?
2. 你申请工作以前会做什么准备?
3. 你滑过滑板吗? 觉得怎么样?
4. 你什么时候照相?
5. 你怎么花你的零用钱?

## Activity 18.25

Translate the following passage into English.

我记得很清楚，我八岁的生日，父母开始给我零用钱。爸爸对我说"这是你的。你可以把钱放在银行里，或者也可以去购物，都行。"从那时候到现在，我很少花钱。我很想去下个月的音乐会。虽然票很贵，可是我的钱够了!

## Activity 18.26

Translate the following passage into Chinese.

I am an only child. Because of this, pressure to get good grades is especially big. My mother won't let me go play with friends; she says studying is most important. She wants me to speak fluent Japanese. She told me that if my Japanese is good, I will certainly find a good job.

# Unit 19 Section 1 Vocabulary

| | | |
|---|---|---|
| 杂志 | zázhì | magazine |
| 深 | shēn | deep |
| 紫色 | zǐ sè | purple |
| 衬衣 | chènyī | shirt |
| 粉红色 | fěnhóng sè | pink |
| 感觉 | gǎnjué | feeling |
| 浅 | qián | light (colour) |
| 橙色 | chéng sè | orange |
| 夹克 | jiākè | jacket |
| 皮带 | pídài | leather belt |
| 美丽 | měilì | beautiful |
| 一般 | yībān | in general |
| 难看 | nánkàn | ugly |
| 领带 | lǐngdài | necktie |
| 大衣 | dàyī | overcoat |
| 流行 | liúxíng | fashionable |
| 眼镜 | yǎnjìng | glasses |
| 世界 | shìjiè | the world |
| 电视剧 | diànshìjù | TV play, soap opera |
| 美洲 | Měizhōu | The Americas |
| 澳洲 | Àozhōu | Australia |
| 亚洲 | Yàzhōu | Asia |
| 欧洲 | Ōuzhōu | Europe |
| 非洲 | Fēizhōu | Africa |

**Online vocabulary**

| | | |
|---|---|---|
| 歌曲 | gēqǔ | song |
| 未婚 | wèihūn | unmarried |

# Unit 19 Section 1 Stroke Order

|  | | | | | | | | | | | | | | |
|---|---|---|---|---|---|---|---|---|---|---|---|---|---|---|
| 杂志 zázhì | ⺄ | 九 | 九 | 卆 | 朵 | 杂 | | | | | | | | |
| 衬衣 chènyī | 丶 | ⺀ | 礻 | 礻 | 衤 | 衤 | 衬 | 衬 | | | | | | |
| 皮带 pídài | 乛 | 厂 | 广 | 皮 | 皮 | | | | | | | | | |
| 美丽 měilì | 一 | 丆 | 丽 | 丽 | 丽 | 丽 | 丽 | | | | | | | |
| 一般 yībān | 丿 | 丿 | 月 | 月 | 舟 | 舟 | 舟 | 舡 | 般 | 般 | | | | |
| 眼睛 yǎnjīng | 丨 | 冂 | 月 | 月 | 目 | 目 | 目 | 目 | 眼 | 眼 | 眼 | | | |
| | 丨 | 冂 | 月 | 月 | 目 | 目 | 目 | 目 | 睛 | 睛 | 睛 | 睛 | 睛 | |
| 电视 diànshì | 丶 | ⺀ | 礻 | 礻 | 礻 | 视 | 视 | 视 | | | | | | |
| 世界 shìjiè | 一 | 十 | 廿 | 廿 | 世 | | | | | | | | | |
| | 丨 | 冂 | 冂 | 田 | 田 | 罒 | 界 | 界 | 界 | | | | | |
| 美洲 Měizhōu | 丶 | 丶 | 氵 | 氵 | 汌 | 洲 | 洲 | 洲 | 洲 | | | | | |
| 演员 yǎnyuán | 丶 | 丶 | 氵 | 氵 | 氵 | 泞 | 泞 | 泞 | 演 | 演 | 演 | 演 | 演 | 演 |
| 医院 yīyuàn | 一 | 丆 | 𠂆 | 匚 | 医 | 医 | 医 | | | | | | | |
| | 阝 | 阝 | 阝 | 阝 | 院 | 院 | 院 | 院 | 院 | | | | | |

# 娱乐媒体 Entertainment and Media

## Section 1 – Celebrities

Activity 19.1

Practise writing the following characters:

杂

衬

皮

丽

般

眼

睛

视

世

界

洲

演

医

院

## Activity 19.2

Draw lines to match words with their simple definitions.

| | |
|---|---|
| 紫色 | 很多人喜欢 |
| 浅 | 很不美丽 |
| 难看 | 漂亮 |
| 美丽 | 一个颜色 |
| 流行 | 深的反义词 |

## Activity 19.3

Split up the following words and write each under the correct heading according to the tone of each character. As an example, 非 has been completed for you.

非洲　感觉　世界　皮带　杂志

| 1st tone | 2nd tone |
|---|---|
| 非 | |
| **3rd tone** | **4th tone** |
| | |

## Activity 19.4

Draw a line to match each Chinese word with its English translation.

| | |
|---|---|
| 领带 | jacket |
| 皮带 | glasses |
| 眼镜 | belt |
| 眼睛 | eyes |
| 夹克 | overcoat |
| 衬衣 | necktie |
| 大衣 | shirt |

## Activity 19.5

Write the pinyin and tones of these sentences.

1. 你更喜欢粉红色还是橙色?

2. 我一般不喜欢看电视剧。看电影更好。

3. 欧洲每年都有很大的演唱比赛。

4. 这是什么歌曲? 好听极了!

5. 你怎么忘了带眼镜呢? 你看得清楚吗?

## Activity 19.6

Below, fill in the blanks with appropriate words.

1. 你去过□□□吗? 我要去那儿旅游。

2. 叔叔在□□工作。他□□以后跟我们见面。

3. 你的□□很□□! 是在哪里买的?

4. 我从来不在书店买□□。□□可以免费看!

美丽　下班　南美洲　医院　杂志　衬衣　网上

## Activity 19.7

Translate the following sentences into Chinese.

1. The day before yesterday I went shopping with my friend. I bought a red leather belt.
2. The basketball team members wear green overcoats. Very ugly!
3. This is the world's prettiest place.
4. In general, he does not read magazines.
5. Do you think this shirt is fashionable?

## Activity 19.8

Match the start and end of each sentence to make a logical complete sentence.

| | |
|---|---|
| 你是商人，应该有 | 你有什么感觉？ |
| 听到好听的歌曲的时候， | 是浅紫色的。 |
| 如果她不戴眼镜， | 可是我自己不要穿 。 |
| 妹妹的鞋子 | 她什么都看不见。 |
| 听说那种夹克很流行， | 很多领带吧。 |

## Activity 19.9

Below are some everyday activities. Using 地, write the manner in which you did each thing yesterday. Feel free to make up responses rather than give real ones.

1. 吃早饭
2. 穿上衣服
3. 走进教室
4. 讲中文
5. 给父亲看数学成绩

Some helpful adverbs:
高兴，怕，慢，快，生气，不好意思，担心，吵，严肃，流利，友好，清楚，忙，快乐

### Activity 19.10
Read the passage about Sonya's dream job and answer the questions in English.

我从小就喜欢表演。我常常唱歌给别人听，还会跳舞。我从来不怕表演。跳舞是我最大的乐趣。我的理想工作是当明星，比如一位职业的演员。可是，我跟父母说了，他们很担心。他们说演员的工作很累，赚的钱也比较少。可是没关系！我很相信自己。我一定会做演员。

1. When Sonya was a little girl, what did she do fearlessly?
2. What was her biggest pleasure?
3. What profession did she dream about?
4. When she spoke to her parents, how did they respond?
5. How did this affect her self-confidence?

### Activity 19.11
Answer the questions for yourself in Chinese. Use complete sentences.

1. 你最想去什么地方旅游？
2. 你喜欢买什么样的衣服？
3. 请说一下你喜欢看的电视剧。

### Activity 19.12
Translate the following passage into English.

上个星期我和哥哥去东南亚音乐会。我们都没听过那种音乐。有一位名的马来西亚歌手来到伦敦表演。他们的音乐很好听。那天晚上我回家以后，觉得很好。可是我不知道亚洲的历史和文化。现在我对亚洲很感兴趣。我明年打算去很多亚洲国家。

歌手 gēshǒu = singer

### Activity 19.13
Translate the following passage into Chinese.

In general, I like buying clothes online. I can often find clothes that are not only fashionable, but also cheap. If the size is wrong, it is easy to return the clothes. Yesterday I bought a black shirt. My younger brother said it was ugly, but I thought it was beautiful.

# Unit 19 Section 2 Vocabulary

| | | |
|---|---|---|
| 动画片 | dònghuà piàn | cartoon |
| 有趣 | yǒu qù | interesting, funny |
| 看小说 | kàn xiǎoshuō | to read a novel |
| 友好 | yǒuhǎo | friendly |
| 亲切 | qīnqiè | amiable |
| 成功 | chénggōng | success |
| 厉害 | lìhai | severe |
| 手表 | shǒubiǎo | wristwatch |
| 不过 | búguò | however |
| 却 | què | yet (unfortunately) |
| 日记 | rìjì | diary |

**Online vocabulary**

| | | |
|---|---|---|
| 差不多 | chàbuduō | almost, about the same |
| 恨 | hèn | to hate |

# Unit 19 Section 2 Stroke Order

亲切
qīnqiè

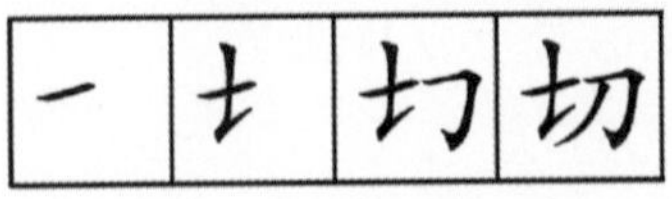

成功
chénggōng

厉害
lìhai

手表
shǒubiǎo

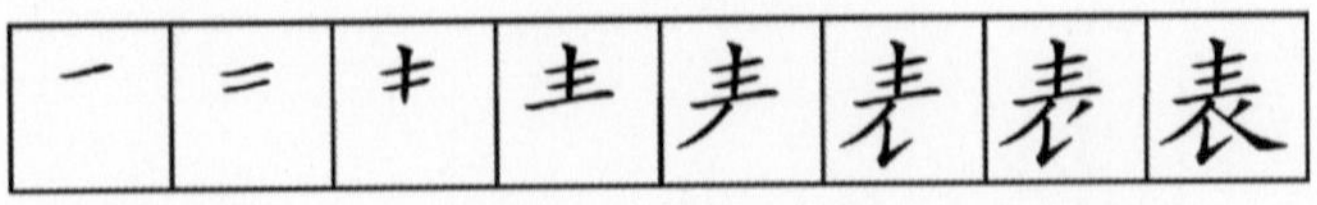

却
què

唱歌
chàng gē

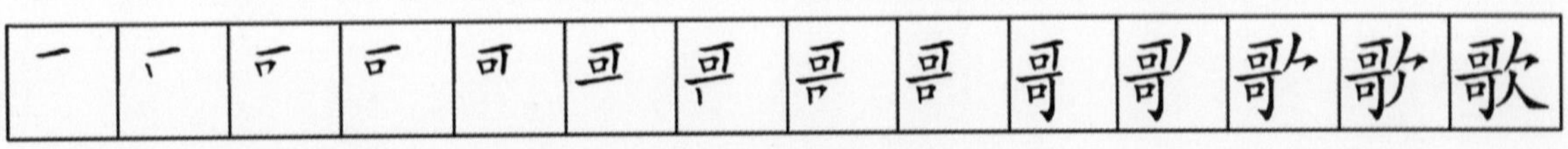

帮助
bāngzhù

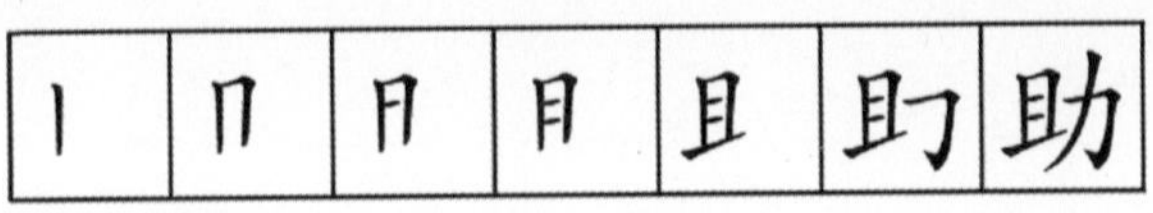

颜色
yánsè

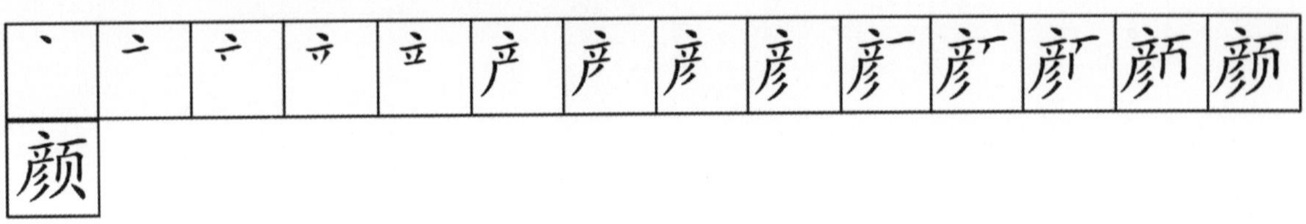

跑
pǎo

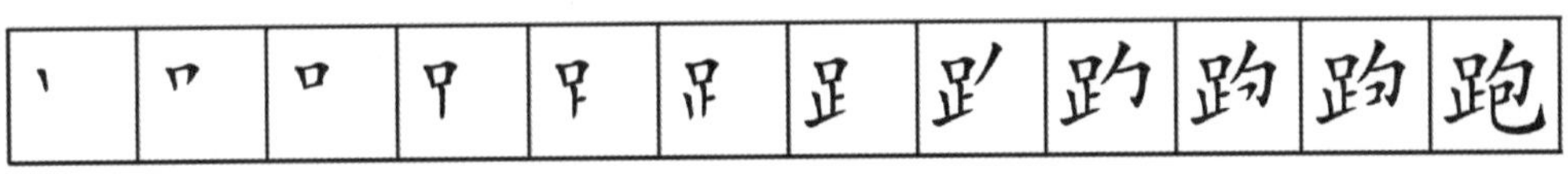

跳舞
tiàowǔ

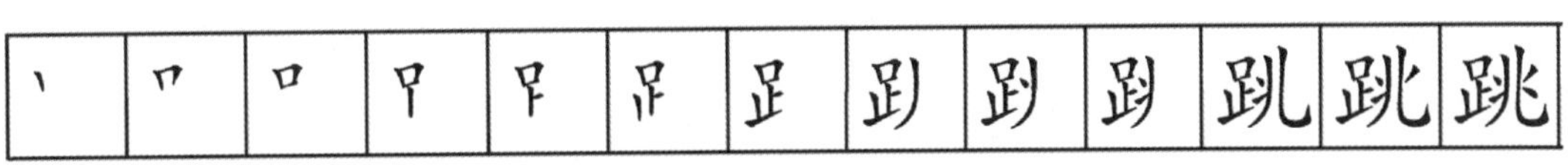

话
huà

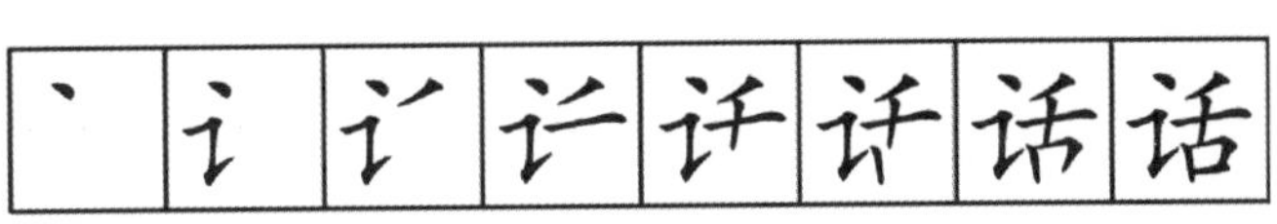

休息
xīuxi

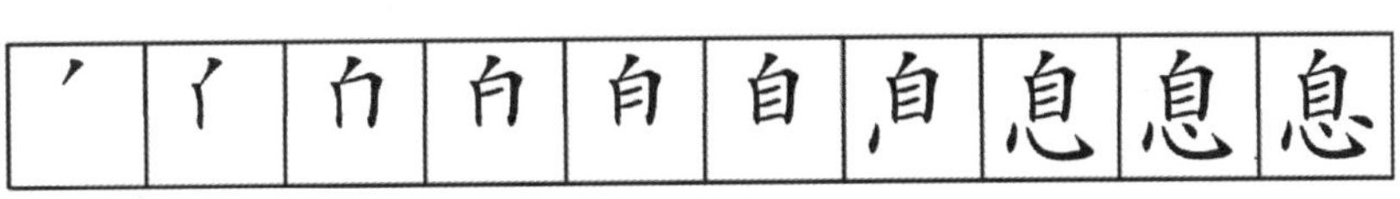

# Section 2 – Inspirational stories

Activity 19.14
Practise writing the following characters:

| | | | | | | | | | | | |
|---|---|---|---|---|---|---|---|---|---|---|---|
| 切 | | | | | | | | | | | |
| 功 | | | | | | | | | | | |
| 历 | | | | | | | | | | | |
| 表 | | | | | | | | | | | |
| 却 | | | | | | | | | | | |
| 唱 | | | | | | | | | | | |
| 歌 | | | | | | | | | | | |
| 帮 | | | | | | | | | | | |
| 助 | | | | | | | | | | | |
| 颜 | | | | | | | | | | | |
| 跑 | | | | | | | | | | | |
| 跳 | | | | | | | | | | | |
| 话 | | | | | | | | | | | |
| 息 | | | | | | | | | | | |

### Activity 19.15
Draw lines to match each word with its simple definition.

| | |
|---|---|
| 厉害 | 有意思 |
| 有趣 | 给你看时间 |
| 恨 | 每天用的本子 |
| 日记 | 讨厌 |
| 手表 | 很严重，很好 |

### Activity 19.16
Split up the following words and write each under the correct heading according to the tone of each character. As an example, 成 has been completed for you.

成功　亲切　友好　小说　有趣

| 1st tone | 2nd tone |
|---|---|
| | 成 |
| **3rd tone** | **4th tone** |
| | |

### Activity 19.17
Draw a line to match each Chinese word with its English translation.

| | |
|---|---|
| 差不多 | interesting, amusing |
| 不过 | yet, however |
| 动画片 | to hate |
| 有趣 | cartoon |
| 恨 | terrible, awesome |
| 亲切 | almost, about the same |
| 厉害 | amiable |

### Activity 19.18
Write the pinyin and tones of these sentences.

1. 他的领带很有趣！
2. 那本小说我看过了，我感觉很好。
3. 你网球打得很厉害！
4. 老师很亲切地说，"今天没有考试。"
5. 爸爸申请了新工作，不过他没成功。

### Activity 19.19
Below, fill in the blanks with appropriate words.

1. 我给女朋友送什么□□呢？给她送□□吧。
2. "来，你用我的□□吧"他□□地说。
3. 我和你的成绩□□□。
4. 你觉得哪个□□好？我自己□□喜欢浅橙色的。

礼物　差不多　比较　手表
颜色　友好　电话

### Activity 19.20
Translate the following sentences into Chinese.

1. She never told me that she loves to sing and dance!

2. He said he writes his diary every day, but I don't believe him.

3. How many novels did you read this year?

4. He runs very well. For example, yesterday he ran 20 kilometres.

5. Best wishes for you to be successful!

### Activity 19.21
Match the start and end of each sentence to make a logical complete sentence.

| | |
|---|---|
| 他很喜欢我的妹妹， | 我要看天气预报。 |
| 我必须准备考试， | 她却不喜欢我。 |
| 我听说明天会变冷， | 可是我没有时间准备。 |
| 弟弟说他没有钱， | 可是我忘了带照相机 。 |
| 我很要拍照片， | 但是他买了一辆新车。 |

### Activity 19.22
Below are some personal attributes or descriptions. Pick a celebrity (or somebody you know) for each and write a sentence stating whether that person does or does not have that trait.

1. 高
2. 友好
3. 严肃
4. 努力
5. 流行

## Activity 19.23

Read the passage about Anton's life and answer the questions in English.

画画是我的乐趣。我小时候一直喜欢看动画片。看了两个小时的电视以后，我妈妈总是对我说"好了！不可以看了！"然后，我会拿着笔和本子出去，在地上坐着画画。有时候我会画自己想到的东西；有时候我画出我看见的东西。我长大以后参加了艺术比赛，拿到了金牌。我也学会了用电脑做动画片。大学毕业以后，我要申请画师的工作。现在，你可以在杂志上看到我的画。

1. How long was Anton allowed to watch TV as a child?
2. What did he like to draw as a child?
3. What did he learn when he got older?
4. What will he do after graduating university?
5. Where can you see his artwork now?

## Activity 19.24

Answer the questions for yourself in Chinese. Use complete sentences.

1. 你写日记吗？在日记写什么事儿？
2. 请说一下你最喜欢的小说。
3. 你帮助过别人吗？请你说一说。

## Activity 19.25

Translate the following passage into English.

去年我阿姨开了一个网上的公司，卖自己做的滑板。虽然她没有工作背景，也没有经验，但是她很有兴趣，也很能干。她的滑板都很美丽。我自己不会滑，不过我的朋友买了她的滑板。我相信她的新公司能成功。

## Activity 19.26

Translate the following passage into Chinese.

My classmate really loves to dance, and he dances really well. Although the amount of time we've been dancing is about the same, he is still much better than me. His ideal is to become a celebrity. I believe he will be successful. I've heard that successful people can make a lot of money.